The Changing Worlds of Geography

A Critical Guide to Concepts and Methods

JAMES BIRD

CLARENDON PRESS · OXFORD
1993

Oxford University Press, Walton Street, Oxford OX2 6DP

Oxford New York Toronto
Delhi Bombay Calcutta Madras Karachi
Kuala Lumpur Singapore Hong Kong Tokyo
Nairobi Dar es Salaam Cape Town
Melbourne Auckland Madrid
and associated companies in
Berlin Ibadan

Oxford is a trade mark of Oxford University Press

Published in the United States by
Oxford University Press, New York

© Oxford University Press 1989, 1993

All rights reserved. No part of this publication may be reproduced,
stored in a retrieval system, or transmitted, in any form or by any means,
without the prior permission in writing of Oxford University Press.
Within the UK, exceptions are allowed in respect of any fair dealing for the
purpose of research or private study, or criticism or review, as permitted
under the Copyright, Designs and Patents Act, 1988, or in the case of
reprographic reproduction in accordance with the terms of the licences
issued by the Copyright Licensing Agency. Enquiries concerning
reproduction outside these terms and in other countries should be
sent to the Rights Department, Oxford University Press,
at the address above

This book is sold subject to the condition that it shall not, by way
of trade or otherwise, be lent, re-sold, hired out or otherwise circulated
without the publisher's prior consent in any form of binding or cover
other than that in which it is published and without a similar condition
including this condition being imposed on the subsequent purchaser

British Library Cataloguing in Publication Data
Data available

Library of Congress Cataloging in Publication Data
Bird, J. H. (James Harold), 1923–
The changing worlds of geography : a critical guide to concepts
and methods / James Bird.
"Second edition"—Pref.
Includes bibliographical references and indexes.
1. Geography—Methodology. I. Title.
G70.B55   1992   910'.01—dc20   92–29463
ISBN 0–19–874182–0 (pbk.: acid-free)

1 3 5 7 9 10 8 6 4 2

Typeset by Colset (Singapore) Ltd

Printed in Great Britain
on acid-free paper by
Biddles Ltd
Guildford & King's Lynn

LANCASTER UNIVERSITY
13 JAN 1994
LIBRARY

DQ
8
B
.93   04974

To Olwen

Change? 'Compare and judge!'
The cries upon my lips;
For the earth way of peoples is rocky,
With many forms and scales; and climes and times;
And their varying forces:
Then cunning are those sciences and
Geo-travellers that choose horses for courses,
And so with fine problems come to grips.

PREFACE

The world about us changes constantly and so do ideas about that world. Constant flux can be both exciting and confusing. So, in the hope of retaining the excitement while dispelling a little of the confusion, here is a guide (not *the* guide) for the young geographer. A geographer is young in spirit if he is not content to cower under the idea that 'geography is what geographers do'; if he remains excited by trying to understand how the discipline has arrived at where it is today; and if he engages in continual speculation about current debates in geography and about where these may be leading us.

Let me explain what prompts this book and what it is about by making six statements:

1. All students of geography in higher education are required to know something about the philosophy and methodology of their subject.

2. 'Student' is a term here used as embracing both the teachers and the taught, but of course they are on different stages of their academic journey.

3. The situation in geography, as in any 'live' academic discipline, is one of many recommended approaches and perspectives. This can be confusing for students of geography, whatever their academic status.

4. This book provides a critical review of many of the approaches and perspectives that have been advocated, generally without acknowledgement of the associated disadvantages. This review is conducted from a particular stance.

5. The stance here advocated is a form of 'critical rationalism' (derived from Popper), within a pragmatic, analytical (combined) methodology-epistemology (PAME), see Chapter 9.

6. Because PAME involves constant-revision of methods used and positions adopted, the book contains an inbuilt invitation to criticize the details of the argument advocated.

My point of view is therefore, on its own terms, ripe for refutation. As a geographer with research interests in human geography, I must try to be fair to physical geography, without the professional competence to satisfy ambition. I am also open to the charge of 'the eye sees only what it seeks', which I find not at all pejorative, but a fact of scientific life: perspective-laden observation. An objective will have been achieved if today's less experienced geographers develop reactively their own points of view, based on their own conjectures, continually tested against evidence in the literature and in the world 'out there'. I do not aim to convert. Experienced geographers will already have devised ways of coping with the many geographical philosophies and methodologies on offer. My arguments may merely serve to strenghten their existing solutions as they gleefully demolish mine, though my experience is that one can learn much from articulating critical comments about the efforts of others.

Where this guide tries to summarize the arguments of whole books and papers, there is a danger that compression will have telescoped stages of those arguments. If such compression also results in difficulty of comprehension, I apologize in advance, although this failing may have the hidden advantage of directing attention to an interesting source. The book is designed to start in a relatively straightforward manner and to deal progressively with more difficult concepts and methods. I am sorry that my own perspective of PAME cannot be explained and justified in a sentence or two. But at least this reflects my view that geography is a deep subject, providing an exciting arena for the contemplation of fundamental philosophical and human problems, a 'thought arena' which is as challenging as that of any other discipline.

So let me guide you on an academic journey through the ever-changing worlds of geography, but I hope no one will trust the guide. Like many other studies of geographical philosophy, this one has yet another -ism hidden within the text: self-justificationism. So I repeat: 'Do not trust the guide!' He may well be in error. Theses herein can be improved upon. *Caveat lector.*

J.B.

Southampton
February 1988

PREFACE TO SECOND EDITION

Recent debates in philosophy prove relevant to topics raised in the text, so I have added an appendix (Appendix I, in two parts). In particular, one reviewer of the first edition thought I ought to have mentioned the work of the philosopher Richard Rorty; so I have provided a short essay (Appendix I, Part II). Debate about Rorty's contribution proves to develop discussion of the 'real world' versus the real world (see pp. 196–9). A second Appendix deals with some of the points raised in other reviews and also in reviews of other recent wide subject areas in geography. All these additions deal with general questions of philosophical approach (though I have tried to concentrate on their relevance for geographers); and so I have kept them separate from the main text.

J.H.B.

Chandler's Ford, nr Southampton
April 1992.

ACKNOWLEDGEMENTS

I should first like to thank my colleagues at the Department of Geography, University of Southampton for the friendly academic atmosphere in which this book was produced, and notably: B. P. Birch, M. J. Clark, K. J. Gregory, B. S. Hoyle, D. A. Pinder, R. J. Small, and J. M. Wagstaff, who each read and commented on a chapter, and sometimes also other sections. P. Boagey, Geography Departmental Librarian, was unfailingly helpful in tracing material. N. Rescher (University of Pittsburgh) kindly commented on the use made in Chapter 9 of material from his *Methodological pragmatism*, 1977. The debt to all the authors briefly quoted, mostly geographers, is obvious, but the following deserve special mention for permission to make longer quotations:

For use in the text. Edward Arnold for permission to quote from E. M. Forster, *Howard's End*, 1910, and from work first published by me in *Progress in Human Geography*; Editions Kister and A. A. Moles for permission to use the 'twenty-one scientific methods' which first appeared in *La création scientifique*, 1957; Routledge and Kegan Paul for material from I. Scheffler, *Four pragmatists*, 1974; the Editor, Institute of British Geographers, for permission to use material from papers by me first published in *Transactions of the Institute of British Geographers*, 6, 1981, 129–51, and 8, 1983, 55–69); the Editor, *Area*, for permission to use material from a paper by me first published in that journal in 1973; the Association of American Geographers for permission to use material from a paper by me first published in *The Professional Geographer* in 1985; M. Willmott for allowing me to benefit from ideas in an unpublished essay on the work of David Harvey. Random House, Inc. Alfred Knopf Inc. for permission to quote from S. Marcus's introduction to *The continental op*, 1974.

For use in figures. Verso (New Left Books) for permission to base a diagram, Fig. 1, on one in P. Feyerabend, *Against method* 1975; J. H. Paterson for a diagram, here redrawn as Fig. 2, the original of which appeared in his *Other laws, other landscapes*, 1975; L. K. Bragaw, H. S. Marcus, G. C. Raffaele, J. R. Townley, and Lexington Books for permission to base Fig. 3 on diagrams in *The challenge of deepwater terminals*, 1975; the Institute of British Geographers for permission to base Figs. 6a and 6b on diagrams by S. A. Schumm in the *Transactions of the*

Institute of British Geographers, 4, 1979, Fig. 13, 501 and Fig. 2, 490 respectively; Pion for permission to redraw Fig. 1 from P. Haggett, *Towards the dynamic analysis of spatial patterns*, 1978, 205–10, see Fig. 6d; Fig. 8 'actualizes' a diagram suggested by the text in P. Jackson and S. J. Smith, *Exploring social geography*, 1984, 4–5, by their permission; Oxford Polytechnic Department of Geography for permission to redraw as Fig. 9 a diagram by R. J. Chorley in D. Pepper and A. Jenkins (eds.), *Proceedings of the 1975 National Conference on Geography in Higher Education*, 1976; the Association of American Geographers for permission to produce Fig. 18 as a redrawn, simplified version of a diagram by Desbarats originally appearing in their *Annals*, 73, 1983, 340–57; the Editor of *Transactions of the Institute of British Geographers* for permission to republish as Fig. 20 a diagram of mine which first appeared in the *Transactions* 6, 1981, 142; the Royal Scottish Geographical Society for permission to republish as Fig. 21 a diagram of mine which first appeared in the *Scottish Geographical Magazine*, 91, 1975, 159; Sir John Eccles and Springer for permission to redraw as Fig. 22 a diagram from K. R. Popper and J. C. Eccles, *The self and its brain*, 1977, 359; Blackwell for permission to redraw as Fig. 24 a diagram, and quote accompanying explanatory text, from N. Rescher, *Methodological pragmatism*, 1977, Fig. 4, 107.

For use in tables. The Editor, *American Economic Review*, for material from a paper by J. S. Bain (44, 1954, 15–39) used as a basis for Table 7; the Editor, *Geografiska Annaler*, for material from a paper by M. K. Watson (60B, 1978, 36–47), used as a basis for Table 8; Edward Arnold for material from R. J. Johnston, *Philosophy and human geography*, 1983, as a basis for Table 9; Methuen for material from R. Peet, introduction to *Radical geography: alternative viewpoints on contemporary social issues*, 1977, used as a basis for Table 13; J. P. Cole for material from his *The poverty of Marxism in contemporary geographical applications and research*, 1986, used as basis for Table 18; Cambridge University Press for material from F. C. Bartlett, Types of imagination, *Journal of Philosophical Studies* (3 1928, 78—85), used as a basis for Table 28; Oxford University Press for material from C. W. Mills, *The sociological imagination*, 1959, as a basis for Table 29. All the figures were drawn in the Cartographic Unit, Department of Geography, University of Southampton.

My grateful thanks are due to Enid Barker and Laurien Berkeley of Oxford University Press for their skill and patience in seeing the book into print.

Finally, my deepest thanks to my dearest wife, Olwen, to whom the book is dedicated, as Secretary to the project, and for whose unfailing help no amount of rhododendrons can recompense.

CONTENTS

xiv   Contents

Figures

Tables

1

Science as Constant-Revision and the Beginning of Modern Geography

Journey together

We shall start by going back to the heady days of the 1960s when a 'new geography' was proclaimed, when at last geographers seemed to reach agreement that their subject could employ scientific methods, although a reaction soon set in (Chapter 1). Geographers are now faced with a bewildering variety of advice on how to practise their art or science and an equally bewildering gamut of scales. Scale is certainly a basic question in geography, and consideration of which strategy to use in exercising scalar choice (Chapter 2) is a good practice run for the problem of choice when confronted with the array of '-isms'. For geography has a flank open to philosophical and methodological developments in all other subjects. So three chapters (3, 4, 5) allow these -isms to pass in review; and I try to give an account of the advantages and disadvantages of each. Though where in the world will you find an unbiased guide? We study 'the world about us', but the actors in that world have their own 'worlds' as perspectives for action. No wonder human geographers have tried to develop a behavioural geography and to deal with some of the pitfalls that have opened up in consequence (Chapter 6). The same applies to geographers' experience of systems, which have proved much more important in physical geography than in human geography (Chapter 7). In attempting creative work in any discipline perhaps the formulation of a problem and then finding a first possible attack on it are the most difficult tasks, and so a variety of creative strategies and tactics is considered, with geographical examples (Chapter 8).

Finally, we see that changes in geography occur in more than one world, and a recommended strategy is argued which, while based on the critical rationalist version of the scientific method, includes modifications so that it is no longer a case of merely adopting a simple natural science model—a procedure which seemed good enough a quarter of a century ago (Chapter 9).

My conjecture, your refutation

What is modern geography? Is it a science? If so, what is the scientific method? These are not only questions but also mighty problems. The first question will not be shirked, but you will have to skip to the last chapter. A basic tenet of this book is that the academic discipline of geography is always an unfinished business, and this even applies to the view adopted of its own history. It follows that this study cannot be a static description of what is in constant flux, but is rather a series of conjectures, with supporting evidence, which you, the reader, can accept or refute in the light of the supporting evidence you can bring to bear. You are in my future, and have the added opportunity of discovering what has happened between when I write and you now read.

Scientific method

There is no agreed description of *the* scientific method. Just imagine the situation if there were. A totalitarian world of procedures would have to be learnt and obeyed. It is difficult to imagine such a universal framework lasting for very long. But that is not to say that there are no rules or tricks of the trade, although such as now seem useful are all on probation. Even this early in our journey, it can be seen that any proposed system must somehow cope with a built-in propensity to be changed, modified, or even refuted. We can begin from a position where there is at least a fair measure of agreement—that the scientific method starts with some kind of problem; and we go so far as to say that problem orientation is the *raison d'être* of scientific enquiry. So problem recognition is a very important part of scientific endeavour, and often a difficult intellectual exercise. One way of identifying a problem is to imagine 'What if?' More frequently problems arise as unexpected consequences of previous endeavours. Having discovered a problem, and wishing to attempt a solution, we must have an idea of how to proceed, and then act on the idea. So we have a simple triad:

Problem ⟶ Idea of a possible solution ⟶ action.

This is so simple and so generalized that it does not get us very far, and so some elaboration is necessary.

The 'action' in scientific method is usually thought of as some form of experiment, and this need not be a physical happening: there are thought experiments. Medawar (1979) presented a classification of four kinds of experiment (Table 1); so let us see where problems (p), ideas (i), and actions (a) appear in the four schemata. In trying to explain a possible description of the scientific method (p), a triad was proposed (i), and a comparison with a classification of experiments (a) reveals that the Galilean type comes closest. The other types of experiment are not discarded, but we have to recognize that 'devising' (Baconian), 'preconceived idea' (Aristotelian), and 'thought experiment' (Kantian) all derive from the results of previous actions—and this leads to the idea of the continuity of scientific endeavour as an open-ended system.

So we arrive at what Medawar has called the hypothetico-deductive method of science based on the sequence presented by Popper (1972, 287) as

$$P_1 \rightarrow TT \rightarrow EE \rightarrow P_2,$$

where P_1 is the initial problem, TT the tentative theory (working hypothesis or trial application of a theory), EE the attempt at error elimination leading to a residual irreducible problem or problems, P_2. But the process is not cyclic, for P_2 is different from P_1. In practice a number of TT are tried in order to help the process of EE. The sequence is patently open-ended; and prescriptive laws and dogmas are avoided because even the most successful trial

TABLE 1.   Medawar's fourfold classification of types of experiment

Baconian	devising happenings (a) and contriving experiences (a) to see what would happen (p)
Aristotelian	an experiment (a) to demonstrate the truth of a preconceived idea (i)
Galilean	an experiment (a) devised as a critical discrimination (i) between possibilities (p)
Kantian	thought (i) experiments (a)

(a) = actions, (p) = problems, (i) = ideas

Source: Medawar (1979, 69–75), with (a), (p), and (i) added.

4 Science as Constant-Revision

solution is still 'on trial', on probation, because a better TT may come along which explains more of the empirical data available (see Fig. 1). Instead of seeking truth, Popper neatly employs the term 'verisimilitude'. So the following widely-held formulation of the scientific method is wrong:

$$P_1 \longrightarrow TS \longrightarrow V \longrightarrow T$$

where V is the validation process of the trial solution leading to a

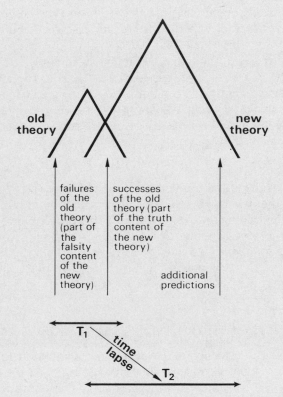

FIG. 1. *Feyerabend's view of Popper's schema for the scientific method*

Note: This has been redrawn from the original in Feyerabend (1975, 174), where it illustrates a by no means favourable discussion of Popper's 'critical rationalism' (Popper's own term), but which, nevertheless, also contains a concise account of the schema being criticized. The diagram has been redrawn by turning it into a mirror image of the original and adding the time notation.

Source: Bird, 1979, 121.

law which is true with a capital T. There are two problems with this schema: we can never be sure that a further observation or instance will not be found which runs counter to the validated solution; second, the sequence is closed, and one can break free only by breaking the conclusion.

The Popperian 'theory on probation' schema has been called the hypothetico-deductive method by Medawar (1969), and he gives this generalized description of it in an elegant passage:

The scientific method is a potentiation of common sense, exercised with a specially firm determination not to persist in error if any exertion of hand or mind can deliver us from it. Like other exploratory processes, it can be resolved into a dialogue between fact and fancy, the actual and the possible; between what could be true and what is in fact the case. The purpose of scientific enquiry is not to compile an inventory of factual information, nor to build up a totalitarian world picture of natural Laws in which every event that is not compulsory is forbidden. We should think of it rather as a logically articulated structure of justifiable beliefs about nature. It begins as a story about a Possible World—a story which we invent and criticize and modify as we go along, so that it ends by being, as nearly as we can make it, a story about real life. (p. 59)

It will be apparent that in this version of the scientific method the appearance of some form of theoretical device appears quite early in the process, in true deductive fashion—the method of reasoning from theory to facts, from the general to the particular. This has been captured by Gombrich (1960, see index) in the phrase, 'making before matching'. Our observations are always theory laden. In fact, Popper has gone further and mounted arguments against induction, where information is first amassed in the hope that a theory or generalization will subsequently arise as a result of repeated instances. The argument denying the logic of induction, moving from the particular to the general, is summarized in Table 2. This argument often meets resistance. Surely, for example, we respond to sense-data inductively? This 'common-sense' view runs up against the belief of many psychologists that even 'perceptions are hypotheses'.

As you might expect Popper's schema of the scientific method has come in for fundamental criticism (see, for examples, Chalmers, 1978; O'Hear, 1980). Another critic, Feyerabend (1975), also gives an account of Popper's 'critical rationalism' (see Fig. 1). For the moment, one can appreciate the cleverness

TABLE 2.   The logical impossibility of induction

---

1   No observation or experiment, however extended, can give more than a finite number of repetitions.

2   Therefore, the statement of a law—B depends on A—always transcends experience.

(Hume's problem of induction (expressed in 1 and 2 above) restated below.)

3   Are we rationally justified in reasoning from instances or from counter-instances of which we have had experience to the truth or falsity of the corresponding laws, or to instances of which we have had no experience?

4   Answer: we are not justified in reasoning from an instance to the truth of the corresponding law. We are justified in reasoning from a counter-instance to the falsity of a universal law (that is, any law of which it is a counter-instance).

= 5   *Induction is logically invalid*, but

6   Refutation or falsification is a valid way of arguing from a single counter-instance to—or, rather, against—the corresponding law.

= 7   The laws that science proposes[a] and uses remain for ever guesses, conjectures, hypotheses.

Solution: Laws in 7, on probation only.

Note: This table deliberately puts matters starkly, and plenty of examples of inductive thinking can be found in geography. But notice that Moss (1977) observes: 'Such [inductive] thinking has been reduced to a minimum in the developed sciences. . . . Deductive strategies . . . [are] a vital, perhaps essential, methodological tool in geography in general and in physical geography in particular' (pp. 25, 39). The argument of this table is discussed *in extenso* in Stove (1982, 52–84).

[a] The deductive element in the hypothetico-deductive method of science.

Source: Popper (1963, ch. 1; 1974, 1013–27).

of the approach, for these critics are doing just what Popper's method requires: testing the very idea, finding out if there is something wrong with it, and trying to suggest something better in a 'we can learn from our mistakes' process. As a reflection of its own internal drive, Popper's 'logic of scientific discovery' can no doubt be improved upon. It derives from a fundamental desire to explain problems in the world of appearances. But, as we shall see, explanation is not the only objective of modern geography.

Science and geography

There remain two of the opening three questions. What is modern geography? Is it a science? Still proceeding to seek answers in reverse order of the questions, we can identify the second question as the philosopher's demarcation problem: how to distinguish science from non-science. This is a problem of definition—a word which means 'setting bounds'. But if science is a process of 'constant-revision', and if the thesis is that science steadily evolves from a pre-scientific or metaphysical phase, the actual process of demarcation itself becomes a problem because we are probably trying to find a marker within a continuum. Catastrophe theory has, however, shown that small continuous changes can produce a sudden flip-over result: it is the last straw that breaks the camel's back. Before looking at the problem with reference to geography, here is Popper's proposal for the demarcation of science from non-science:

But there is another, a special kind of boldness—*the boldness of predicting* aspects of the world of appearance which so far have been overlooked but which it must possess if the conjectured reality is (more or less) right, if the explanatory hypotheses are (approximately) true. It is this more special kind of boldness which I have usually in mind when I speak of bold scientific conjectures. It is the boldness of a conjecture which takes a real risk—the risk of being tested, and refuted; the risk of clashing with reality.

Thus my proposal was, and is, that it is this second boldness, together with the readiness to look out for tests and refutations, which distinguishes 'empirical' science from non-science, and especially from pre-scientific myths and metaphysics. (Popper, 1974, 980–1)

This demarcation proposal is consistent with Popper's idea of the scientific method as based on conjectures and refutations. But this method of demarcation hardly yields a date on which geography changed from a pre-science to a science. And remember, this flip-over is the only Revolution with a capital R that a discipline ever undergoes; once it begins to use scientific methods, then it is a case of revolution in permanence.

The demarcation between science and non-science can be looked at in another way. We can regard the TT phase of the scientific method as a striving for generalization; science is certainly the enemy of the unique. Now if an academic discipline is believed to be based on the fundamental uniqueness of its objects of study, then the scientific method cannot be applied. An idiographic approach has run as a thread through geography. And generations of geography students have had to struggle with the idiographic/nomothetic polarities. A powerful expression of the idiographic perspective was the theme of geography as areal differentiation. This appeared in the great monograph of Hartshorne (1939, 244 and 463), although the penultimate section was entitled: 'What kind of science is geography?' We may call this reference Hartshorne I, because he changed his view later when he came to write a perspective on the nature of geography (Hartshorne, 1959), which we can call Hartshorne II. Hartshorne I was the only thoroughly argued monograph on the nature of geography available in English after the Second World War. A later British effort by Wooldridge and East (1958) was less Olympian in scope but very influential in its day. It contained the following two statements:

. . . what interests the professional geographer . . . is the essential pattern and quality of the earth's surface—'places' or 'areas' and the great difference between them. In simplest essence the geographical problem is how and why does one part of the earth differ from another. (p. 28)

In the ugly and unfamiliar jargon of the philosophers who classify knowledge, it [geography] attains to little nomothetic quality. Geography is essentially idiographic. It is for this reason that, in the narrower sense of the term, it is denied the title of science. (p. 145)

At least they realized that an idiographic view was non-scientific, or, in the light of what happened later, pre-scientific. It is true that in their conclusion they stated that geographers should attempt generalizations, but, because they advocated that these should be based on patient, detailed work, they were advocating the inductive method (p. 173).

At some time during the following decade, from 1958, geographers came to believe that their subject had undergone a quantitative revolution, and that somehow this made the subject more scientific. This ignored the fact that a change from pre-science to science is a philosophical revolution and not based on a developing assistant technology. The paper which gave rise to the widespread belief in the quantitative revolution was Burton (1963), and the term seemed reinforced by the increasing use of statistics in geography, the translation of statistical techniques from other disciplines, and the steady growth and sophistication of computation facilities. The term 'revolution' implies something dramatic acting over a short span, and therefore relatively easy to date. But it is difficult to imagine finding a date or short period before which quantitative methods are absent from the discipline and after which they are all-pervasive. Even if it could be argued that a school of geographers suddenly saw the blinding advantages of quantification overnight, this is not a conceptual revolution but an enabling change. Sciences are not distinguished from non-sciences by the amount of quantification they use, but by whether or not they employ the hypothetico-deductive method. And this cannot be employed if the data in question have fundamental qualities of uniqueness.

Yet a careful reading of Burton's quantitative revolution founding paper hardly supports the great edifice of revolutionary thought. For a start, the paper is entitled 'The quantitative revolution and theoretical geography'. He certainly dated a quantitative revolution which reached 'its culmination in the period from 1957 to 1960, and is now over. . . . *An intellectual revolution is over* when the revolutionary ideas themselves become part of the conventional wisdom' (pp. 152–3). But previously his argument had weakened this idea of a quick change from a quality-dominated discipline to one completely oriented to quantification: 'Some scholars have chosen to regard the revolution in terms of a qualitative–quantitative dichotomy. It does not help to cast the debate in this form' (p. 151). This seems to undermine the whole concept of a quantitative revolution, and there is confirmation in the next paragraph: 'The quantity–quality debate has also been allowed to embrace and perhaps conceal a number of related but distinct questions'.

Burton then lists a number of polarities, and it is instructive to consider these one by one (see Table 3). The hindsight comments may seem hard on Burton, but there is no mistaking his rightness in attempting to downgrade the quantitative revolution in favour of the only real revolution—from a pre-science to a science. Lest this seem a wilful misreading of a historically important paper, let us end the commentary on it with one last quotation: 'The desire to avoid . . . confusion reinforces my inclination to sidestep the quality–quantity issue, and to view the movement toward quantification as a part of the general spread and growth of scientific analysis into a world dominated by a concern with the exceptional

TABLE 3. Burton's polarities (1963) and some comments

Polarities	Comments
measurement by instruments versus direct sense data	This is a quantity–quality polarity.
rational analysis versus intuitive perception	What is meant by 'rational analysis'? Is this the scientific method? If so, what version of the scientific method?
cold and barren scientific constructs versus rich variety of daily sense-experience	Presumably the terminology is ironic. Appears to be deduction versus 'common-sense' induction.
continuously varying phenomena versus discrete cases	Nomothetic applicability (with stochastic possibilities) versus idiographic classification.
nomothetic versus ideographic [sic]	The above now explicitly expressed.

Source: Burton (1963, 151).

Four definitions, with etymology:

idiographic 'adj. . . . [*idio* (distinct) + *graphic* (to write) . . .]: relating to, involving, or dealing with the concrete, individual, or unique . . . contrasted with nomothetic' (*Webster's Third New International Dictionary*, 1961).

nomothetic 'adj. . . . [. . . from *nomo-* [usage, custom, law] + *-thetēs* one who establishes . . .] relating to, involving or dealing with the abstract, recurrent, universal: formulating general statements or scientific laws' (*Webster's Third New International Dictionary*, 1961). 'That pertains to or is concerned with the study or discovery of general (scientific) laws, esp. as contrasted with idiographic study' (*A Supplement to the Oxford English Dictionary*, 1986).

nomology '[n. nomological, adj.] [*nomo* + *-logy* (word, reason, speech, account)] the science of the laws of the mind' (*Webster's Third New International Dictionary*, 1961).

stochastic 'adj. . . . [Gk. *stochastikos* (skillful in aiming, proceeding by guesswork) . . .] (*Webster's Third New International Dictionary*, 1961). 'Randomly determined: that follows some random probability distribution or pattern, so that its behaviour may be analysed statistically but not predicted precisely' (*A Supplement to the Oxford English Dictionary*, 1986).

and the unique' (pp. 151–2). Amen to these sentiments! But in 1963 had geography escaped from being 'dominated by a concern with the exceptional and the unique'?

Ten years earlier Schaefer (1953) had dubbed as 'exceptionalist' any belief that geography's scientific methodology was in some way different from that practised generally in science. One of the principal reasons for holding an exceptionalist view was the uniqueness of the location of data arranged in space: 'the degree to which phenomena are unique is not only greater in geography than in many other sciences, but the unique is of the very first practical importance' (Hartshorne, 1939, 432). Schaefer commented (1953, 239, based on Hartshorne, 1939, 432): 'Hence generalizations in the form of laws are useless, if not impossible, and any prediction in geography is of insignificant value.'

Schaefer died young; his paper was published posthumously. Hartshorne immediately replied (1954, 1955) and went on to publish a book-length perspective on his original monograph (1959; see the discussion in Wood, 1982, and May, 1982). Paterson (1970) remembers hearing Schaefer lecture in the autumn of 1949 at Madison, Wisconsin:

After describing the efforts which he [Schaefer] and some colleagues were making to set up type regions—we should now call them models rather than study regions, he encapsulated his message in words which were a foretaste of much that was to come in succeeding years: 'We are tired of facts, facts, facts. We would rather have laws, laws, laws.' His audience, geographers to a man, gave every evidence of dissent. (p. 6)

These were indeed brave exhortations, but the reef of the belief in the uniqueness of locations lay in the way, blocking geography's change from a pre-science to a science.

The one and only revolution in geography—June 1966

Before 1966 there were many indications that geography was moving away from an exceptionalist position to embrace the general methods of the hypothetico-deductive schema. An event in the literature enables us to date the last straw, when the last idiographic bastion in geography was overthrown—the destruction of the idea that locations could never be anything but unique. The event occurred as a result of an attempt to deal with a problem in regionalization encountered by Grigg (1965).

1. It can be argued that all parts of the earth's surface are unique; classification and regionalization obscure this fundamental fact.

2. Even if this objection is overcome it can be argued that parts of the earth's surface all have the property of location; all locations are unique by definition, hence this is a property that cannot be used as a differentiating characteristic. But a geographical classification which neglects location is of limited value. . . .

All locations are unique by definition; hence, whereas location can be a property, it cannot be a differentiating characteristic, . . . (pp. 476–7)

The point is not whether or not this argument is right or wrong, but that, if it is held that geography includes the study of the location of its data and that 'locations are unique', then geography cannot fully employ the scientific method.

Nine months later, in June 1966, Bunge published his 1½-page commentary on Grigg's paper, asserting that 'Locations are not unique', but insisting they were general. He argued that locations are comparable, witness such terms as 'near', 'far', 'close', 'distant', and 'adjacent', which describe the relativity of locations. While acknowledging the point that Grigg had stated, and which is obvious, that no location is exactly like any other, Bunge went on:

. . . but then no two anythings in the real world are exactly alike. To admit that no two objects are identical, that is, have everything in common, in no way contradicts the statement that these objects may have much in common. . . . Why are geographers so stuck over this simple logical point?

. . . science is the deadly enemy of uniqueness. As the masterful Schaefer taught us, generality is science's weapon in our unending reduction of uniqueness. (Bunge, 1966b, 375–6)

In his reply Grigg (1966) returned to his problems when comparing classification and regionalization procedures, arguing that a point on the earth's surface must be unique. But he did concede that 'locations are essentially relative'. Here was encapsulated the one and only Revolution in geography, because the data studied were henceforth not held to be always discretely differentiated in space and so possessed of unavoidable qualities of total uniqueness. This is not to say that scientific methodology would commandeer the discipline with undisputed sway for ever after. But even proponents of other -isms might concede that June 1966 was the apogee of the view that geography could employ the methods of science because there was no longer any inherent quality in its data that prevented it from so doing. This viewpoint, widely held in the late 1960s and 1970s, gradually came under criticism when it was said that scientific methods were inadequate to confront the full range of problems in geography. Even Bunge modified his position (1974), but his 1966 commentary had, nevertheless, symbolized the one and only revolutionary change, for henceforth it would be a case of constant-revision.

Another view of scientific development, and the reaction against science in geography

All 'new' geographers seem to agree that they are involved in a revolution although the 'type' may vary, e.g. quantitative revolution (Burton, 1963), methodological revolution (Gould, 1969), conceptual revolution (Davies, 1972), statistical and 'models' revolution (Wilson, 1972), plus of course the subsequent 'behavioural' and 'radical' revolutions which are sometimes identified. Seven revolutions in one generation, makes geography 'the Latin America' of the scientific community! (P.J. Taylor, 1976, 141)

The above quotation is totally at variance with the account so far given here. I did comment on this passage a year later by saying that perhaps so many revolutions in so short a time indicate in themselves rather a continuously rolling programme, or something basically wrong with the overturning metaphor (Bird, 1977b, 105). But you might wonder how Taylor and I could come to such contrasting conclusions about the scientific development of geography. A clue is given in the first sentence of Taylor's abstract: 'This paper describes recent trends in British geography from a sociological perspective.' The question now arises as to whether evidence for changes in an academic discipline derives from the data of published material or from within the sociology of the practitioners. These alternatives are summed up in the contrasting viewpoints of Karl Popper and Thomas Kuhn. Kuhn's major work on the philosophy of science, *The structure of scientific revolutions* (1962), has been very influential throughout science. One of the key concepts of the book is that of the paradigm, which, although used by Kuhn in many different ways (see Masterman, 1970), can mean a shared philosophy among a group of scholars, a metatheory guiding their work, which can therefore be described as 'normal science'. We can trace the passage where Kuhn's concept entered geography:

Paradigms may be regarded as stable patterns of scientific activity. They are in a sense large-scale models, but differ from models in that: (1) they are rarely so specifically formulated; and (2) they refer to patterns of searching the real world rather than to the real world itself. Scientists whose research is based on shared paradigms are committed to the same problems, rules and standards, i.e. they form a continuing community devoted to a particular research tradition. In a sense then, paradigms may be regarded as 'super models' in which the smaller scale models are set. (Haggett and Chorley, 1967, 26)

The above passage refers to only one meaning of paradigm that Kuhn intended. He also used the word to mean 'exemplars', referring to standard examples which explain formerly unexplained events in a discipline's agenda. As Mair (1986) points out, all the discussions of paradigm in geography have referred to the first of the above meanings—the metaparadigm idea. The discussion that follows is critical of the metaparadigm concept, but Kuhn's other ideas concerning analogies as epistemology and incommensurability will be referred to in the last chapter.

Two confrontations in print between Kuhn and Popper have been contrived, with Kuhn having the last word in 1970 and Popper having the last word in 1974 (Lakatos and Musgrave (eds.), 1970; Schilpp (ed.), 1974, ii. 798–819). Kuhn's basic argument is contained within the same paper on each occasion, and its very title might be 'Popper versus Kuhn' instead of its actual title: 'The logic of discovery or the psychology of research?' Popper and Kuhn each acknowledge that they have learnt from the other. There is some common ground, but there are basic differences, possibly due to looking at the same material through different spectacles, according to Popper; while Kuhn believes that a 'Gestalt switch' divides one view from the other (Schilpp, 1974, ii. 816 and 1145). This derives from the Gestalt school of psychology, in revolt from an atomistic perspective, in which Gestalten are wholes or frames of reference via which we perceive. Thus a 'Gestalt switch' occurs in optical illusions when we 'see' two images in the same picture by 'willing' a different whole, e.g. the 'young lady or old lady' picture (illustrated by Kirk, 1963, Fig. 4, 365, and n. 17).

A fundamental difference is Kuhn's belief in 'normal' science as the basic state of science. Popper acknowledges that Kuhn has opened his eyes to the existence of 'normal' science, only to deplore it as a 'danger to science' (Popper, 1970). Normal science deals with three classes of problem according to Kuhn: 'determination of significant fact, matching of facts with theory, and articulation of theory' (Kuhn, 1962, 34). He adds that it is a striking feature of these 'normal' research problems how little they aim to produce major novelties, conceptual or phenomenal (ibid. 35). A crisis is the pre-condition for the emergence of novel theories, via 'extraordinary' research leading to a new metaparadigm via scientific revolution (ibid. 77–91).

A criticism of Kuhn is implied by asking what are the scientific criteria for change from belief in one metaparadigm to another? Lakatos (1970, 178) has suggested the 'bandwagon effect' and 'mob psychology' as possible answers. Criticisms that have been levelled against Kuhn's system are conveniently collated in Barbour (1974, 106–8), who goes on to discuss Kuhn's modifications to his original monograph, but there are still no automatic rules for metaparadigm choice (ibid. 111). In contrast to the open-ended nature of the Popper philosophy 'normal science' appears closed—'a closed society of closed minds' (Watkins, 1970, 27). Popper has dubbed 'normal science' as *The Myth of the Framework*: 'if we try, we can break out of our framework at any time. Admittedly, we shall find ourselves again in a framework, but it will be a better and a roomier one; and we can at any moment break out of it again' (1970, 56). Here is the basic difference between Popper and Kuhn encapsulated even in that adverbial phrase 'at any moment'; because if we follow Popper, we enter an open-ended sequence. Harvey (1969a, 486) chose one key stage of this sequence when he asserted, 'By our theories you shall know us.' Perhaps even more important is what lies before us, and our banner should read: 'By the problems we have identified, you shall know us.'

Paterson's concise schema of the chronological development of geography (Fig. 2) shows each stage growing out of the preceding. From the problems of understanding the contemporary world (1) came a desire to look at past worlds (2); and this led to an interest in the processual links between them. Then geographers were tempted to allow the studied processes to run on into the future (4). The predicted geographies are the geographer's perceived

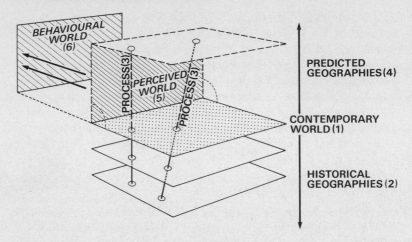

FIG. 2. *Paterson's schema of the chronological development of the field of geography*

Note: Redrawn from the original in J. H. Paterson (1975), by permission of the author. He calls this 'the enlarging content-dimensions of the field' (p. 3), and it is not difficult to see each arbitrary stage as developing out of the problems thrown up by the preceding stage, in open-ended fashion.

future world, but there are other perceived worlds (5), and the geographer is then led to study the behavioural reasons for their existence (6). A shorthand version of this sequence is to say that geography has concentrated on asking the following dominant questions in this order: where? what? when? how? why? and how ought?; each one is superadded to the others (Hart, 1984, Fig. 1, 15).

The success of the scientific method depends on its ability to satisfy our basic curiosity and propensity for attempts to surpass what has been done. Those who have promoted the very permissive nature of the conjecture and refutation process, provided that it is conducted with scrupulous regard to correspondence with the available evidence, perhaps did not foresee that some scientists would still like to grasp certainties such as 'truth', 'verification', and 'final solutions'. Despite their very great differences, Feyerabend, Kuhn, Lakatos, and Popper, have been lumped together as 'four irrationalists' (Stove, 1982) who have promoted a ' "critical attitude" into a categorical imperative of intellectual life' (p. 99). An attack has also been mounted on these same philosophers of science in the journal *Nature* (Theocharis and Psimopoulos, 1987), which provoked vigorous subsequent correspondence, mostly objecting to the attack (*Nature*, 1987, 308 and 689–90; 1988, 129–30). It is interesting to observe how opponents of conjecture and refutation in the hypothetico-deductive method of science and philosophy soon fall into a totalitarian world:

... practitioners of these disciplines [should] stop running down their own professions and start pleading the cause of science and philosophy correctly. This should best be done, first by thoroughly refuting the erroneous and harmful antitheses [as propounded by Feyerabend, Kuhn, Lakatos, and Popper]; secondly by putting forth adequate definitions of such fundamental concepts as objectivity, truth, rationality and the scientific method; and thirdly by putting the latter judiciously into fruitful practice. Only then will the expounding of the positive virtues of science and philosophy carry conviction. (Theocharis and Psimopoulos, 1987, 598)

Who says when the cause of science and philosophy has been pleaded correctly? Who is to judge the adequacy of proposed definitions of objectivity, truth, rationality, and the scientific method? Presumably, the authors of the above quotation have suggestions for answers, but based on what? And if they could supply answers, agreed by other scientists, would they all then forever hug the fetters they had forged?

We can also see attacks on science, however defined, developing in geography (e.g. Zelinsky, 1975), and defences being mounted (Hay, 1979). In addition to such general debates, frameworks for research endeavours in geography are promoted which are very different from any perspective that could be labelled as scientific method; and, moreover, these approaches are very different from each other. For some geographers, this is intolerable pluralism; others seek accommodations via eclecticism. No wonder a guide is needed. But for any guide to be objective is impossible, because he will be seeing the evidence through his own conceptual framework. Johnston (1979*b*), in the preface to his survey of Anglo-American human geography since 1945, agrees, but he does go on to write: 'But although not objective, nor intended to be, the book is neutral. My own opinions are not stated, and are not intendedly implied in anything that has been written (though some of them may be identifiable). There is no commentary, only a presentation of what I perceive to be the salient features' (p. 3). Here we differ. This guide is polemical. But before taking our framework of reference to confront various -isms in geography, it will perhaps be useful to see if it is possible and academically respectable to operate in more than one dimension and at more than one scale.

2

Problems of Scale

Bridge passage

The eyes of a geographer light up when scale is mentioned, because an ability to slip smoothly between scales and compare different levels of approach prove to be of great advantage in practising the discipline. 'Scale' is a word with more than one meaning. For the moment we can build out from the scientific method the basic fact that this type of approach must deal with more than one case or no generalizing statements are possible. Here is the first hint that the hypothetico-deductive method might not serve geography comprehensively, because sometimes problem-solving may require a focus on the understanding of a single event with the generalizing phase consisting of the consideration of the context and consequences of that event. On the other hand, science does not necessarily require the Aristotelian scale of the comprehensibility of its general statements over all the data, although this may certainly remain as an aim. Comprehensibility would require that all extant knowledge in a subject would either by contained in its logical foundation or be deducible from it. For example, physical geography has been defined '*sensu stricto* . . . in terms of the physical processes and organic materials of our environment' (Orme, 1985, 259). But a counterpart generalized statement of this nature is hard to find for human geography and the social sciences generally. A debate on this very problem once occurred in economics as to whether or not that discipline is a theoretical science, and a brief report will indicate that the problem of comprehensibility loomed large, as of course it does for human geography.

Georgescu-Roegen (1966, 109–10) took the view that because economic theory as we know it applies to a particular institutional setting, it does not have comprehensibility and could not be classed as a theoretical science. He also believed that there was a fundamental difference between physical sciences and social sciences (and this of course would separate physical geography and human geography). It was held that, whereas models serve didactic purposes in both areas, in physical science a model is also an accurate blueprint; in social science it is best regarded as an analytical simile (ibid. 116). Hildreth (1967) commented on Georgescu-Roegen's position. After rejecting the Aristotelian concept of a theoretical science, Hildreth asked whether a science was not to be regarded as theoretical if it was forced to employ, temporarily one hopes, several logical structures to interpret relevant observed phenomena, and if its logical foundation was continuously under construction. Hildreth went on to point out that no science can in fact meet Georgescu-Roegen's criteria. He conceded that economic theory as developed in Western countries has aspects which do not apply to other institutional settings, but the problem is no different from that in physical sciences where the question concerns the applicability of laws to other galaxies, and even at the subatomic level in this galaxy. We can further relax the idea of the universal prescriptive law. Not only need it not, temporarily one hopes, cover all the data the discipline can muster, but if we say that the law is stochastic (see Table 3) over a number of cases, rather than causal in every case, tempering certainty with probability, then we can see that human geographers could employ nomothetic methods (Bishop, 1978).

In summary, it might be added that in physical geography testable models can approach an exactness more closely over a greater number of cases than in human geography. But the ambition is the same: comprehensive exact theories to fit all cases, which is impossible. The degree to which this unattainable goal is reached places sciences on a scalar continuum not merely as a result of the prowess of the practitioners of a subject but also as a result of the nature of the data in question: combinations of physical and social variables might be expected to be less tractable to the scientific method than physical variables alone. As this degree of intractability increases, no wonder would-be problem-solvers have sought strategies other than the hypothetico-deductive method.

TABLE 4.   Simple conventions about scale

Level	Typical earth surface processes[a]	Suggested areal extent in hydrological studies[b]	Appropriate level of human spatial organization
macroscale	solar radiation, water resources, flood problems	national or multi-catchment (> 1300 sq. km.)	international,[c] national[c]
mesoscale[d]	tectonic activity, eustatic movement, cyclonic activity, water balance, drainage basin dynamics	from large catchments (1300–25 sq. km.) to small catchments (< 25 sq. km.)	intranational,[c] regional
microscale	tidal currents, frost action, soil-creep, stream-flow	site (water-gauging)	lowest order of administrative unit

Note: See also the relationship of areal studies to a mathematically derived 'G-scale', based on successive logarithmic subdivision of the earth's surface, an index that has never caught on (Haggett *et al.*, 1965).

[a] Examples from Harvey (1968, 72), except those in hydrology, which are from Slaymaker (1968, 72).

[b] Slaymaker (1968, Fig. V.1, 68): 'We have found that the factors affecting the slope of stream channels, and thereby the landscapes of surrounding areas, have markedly different strengths at different spatial scales of analysis' (Penning-Rowsell and Townshend, 1978, 413).

[c] A threefold scalar division common in political geography, and, surprisingly, this has political overtones (see section of text headed 'The politics of scale').

[d] Harvey (1969*a*, 484) sees the meso- (regional) scale as the typical resolution level for geographical work because 'Any phenomena that exhibits significant variation at that resolution level is likely to be the subject of investigation by the geographer.'

A few scalar basics

As geographers we should be nervous about using expressions like 'large-scale' and 'small-scale', because large-scale map sheets cover small areas, and to map phenomena on a small scale is to cover a wide area. The convention in Table 4 seems safer. It was, I believe, Harvey (1968) who first pointed out to geographers the connection between randomness and the appropriateness of various scales:

It is self evident, of course, that if we are seeking to explain micro-variations in pattern, then the relevant processes are different from those that we would consider in seeking to explain macro-variations. . . . Failure to meet random expectation indicates non-randomness in the pattern, but the successful fit of the random expectation model cannot be used to infer that the pattern is random at all scales. . . . If we can identify that scale at which a particular pattern deviates most from randomness, then we may use this knowledge to gain a deeper insight into the structure of that pattern and perhaps some clue as to the manner in which scale and process are interrelated. (pp. 72–4)

The problem presented by induction suggests that in classifying data it is better to go from the general (macroscale) towards the particular (microscale) via a process of logical division, rather than begin by aggregating individuals and proceeding 'up the hierarchy'. Table 5 indicates that this preference has advantages both in computational time and in stability of the classification produced.

Different scales of approach may eventuate in different results

This looks very serious. So some examples ought to be provided. First, a very familiar scalar conflict is between national economic growth objectives and degradation of local environments. The national efficiency requirement for a fast route from A to C may involve disadvantages of noise, congestion, pollution, and sheer alienation of local sites near intervening place B. This is often presented as a conflict between national needs and local environmental protection. Bragaw *et al.* (1975) presented a revised view of such an economy versus environment dichotomy (Fig. 3). Their suggestion looks like a mere rotation of dimensions, but it

TABLE 5.   Fourfold classification of classification procedures

Method	Uses all data	Stability of classification	Relatively shorter computational time?
DP	✓	✓	—
DM	✓	—	✓
AP	—	✓	—
AM	—	—	—

Notes: The checks indicate that the method in question has that columnar characteristic. The AP method does not use all the available data unless there is feedback for amendment after the hierarchy has been built up once; and then of course the original classification is no longer stable.

The two methods of classification are discussed in detail for geographers in Johnston (1976) under the terms 'agglomerative methods' and 'divisive methods'.

D = Method of classification by logical *division* of the general to obtain the particular = *down the hierarchy*.

P = Polythetic, based on more than one characteristic.

M = Monothetic, based on one key characteristic. The lack of stability results from the fact that the classification changes with choice of key characteristic.

A = *Additive* method of classification combining the particulars to reach greater levels of generalization = *up the hierarchy*.

Source: Classification of classification procedures based on a suggestion by J. M. Lambert, oral communication.

suggests that there are more complicated dichotomies of scale. For example, at the microscale both environmental and economic factors are present. At the national scale, optimum growth objectives prompt the question: growth for what? Part of the answer must lie in the quality of life enjoyed by all the citizens at the microscale. So the stark confrontation between economic efficiency at one scale and environmental quality at the other has to be buffered by the political dimension of current discussion within a majority-agreed legal framework. On the one hand, national development is necessary if the country is not to stagnate and thereby lose ground in competition with others. On the other hand, the local quality of life can suffer from the problem of irreversibility. If development is permitted and proves to be a mistake, it usually cannot be undone. The conflict of scales leads into deep waters at the forefront of political discussion.

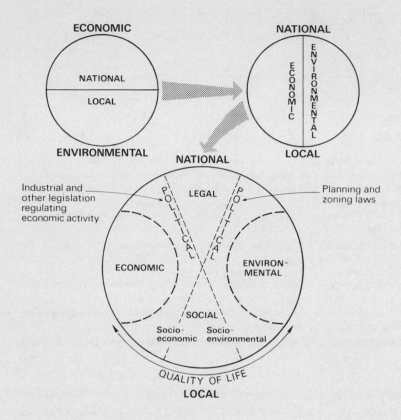

FIG. 3. *Dissolving the dichotomy between national and local scales leading to dimensions of possible compromises*

Note: Redrawn by permission from Bragaw *et al.* (1975, Figs. 5–1, 5–2, and 5–3), where the context is a national economic requirement for deep-water oil terminals in the face of opposition from local environmental pressure groups.

There is a real danger of inferring that statistical patterns at the macroscale are replicated at the meso- and microscales; and this has become known as the 'ecological fallacy'. 'Ecological' here refers to correlations at the macro- and mesoscale. Suppose that in the entire US 50 per cent of both the black and the non-black population vote Democrat and 50 per cent vote Republican, and that 30 per cent of the voters are black, thus:

% of total voters	Democrat	Republican	Total
Black	15	15	30
Non-black	35	35	70
TOTAL	50	50	100

In a southern state two-thirds of the black vote might be Democrat, but this might be counteracted by an increased non-black Republican vote compared with the national average, thus:

% of total voters	Democrat	Republican	Total
Black	20	10	30
Non-black	30	40	70
TOTAL	50	50	100

The same totals for rows and columns appear, and, if only these are available, there is a danger of inferring that they are made up of exactly the same components (for black and non-black voters) at both macroscale (nation-state) and mesoscale (individual states). Johnston (1981, 89, with another worked example) warns that all techniques that are based on correlation coefficients face this issue:

Despite the nature of their data [aggregate data, such as those published in censuses], geographers often wish to infer characteristics of individuals from ecological correlations . . . so that the dangers of the ecological fallacy, of wrongly deducing conclusions about individuals from grouped data, are often severe. (p. 89).

The following three examples deal with clashes between meso- and microscales of approach:

'. . . On small experimental watersheds one can show clearly the effect of differing land uses by comparing an unused control with a watershed which is altered. But if you have a sample of . . . [less than 4 ha], where this would be clearly demonstrated, how do you relate that to something of . . . [25 km², or 400 km², or 2500 km²]?—you cannot apply unadjusted the results from a small watershed to a larger one.' [Leopold, 1960] Leopold is referring particularly here to the difficulty of generalizing much of the data derived from experimental plots of the United States Soil Conservation Service in relation to sediment yield . . . (Slaymaker, 1968, 69).

Next, take two regions A and B. Suppose a study of them comes to the conclusion that their physical and human geographies are more alike than different, as a result of evidence gathered over the whole mesoscale. Now suppose we take two small areas of the same size within each region and compare them. From our regional knowledge we can certainly choose two areas which are a priori likely to give the best chances of 'being more alike than different'. This procedure was once carried out using the South-western Peninsula of England, and Brittany, the north-western peninsula of France (Bird, 1956). The two peninsulas were demonstrated as more alike than different at the mesoscale, but the carefully selected small areas within each peninsula proved to have characteristics very different from each other. The answer to this paradoxical result is that as the scale of study advances towards the meso- and macroscales more generalization becomes necessary, involving the judicious selection and omission of detail. This judgement cannot result in accurate description of every small area that nests within the larger. The regional geographer is like a painter or novelist who must selectively interpret a vast and complicated landscape. The layman, or field-work beginner, is bewildered by the overwhelming detail, for

at each step one sees in the landscape effects of light and shadow which seem to upset the general picture, and of which the layman is in the habit of saying: 'How would a painter interpret that?' To which the artist would reply: 'By not interpreting it.' (Translated from George Sand, 1861, 204; see also Kennedy, 1977, 156 on the 'scale-dependence of empiric generalizations')

A final example of the 'clashing of scales' is where case-study work (empirical investigation) operates at the microscale and theory construction at the meso- and macroscales (see Table 8). N. Smith (1987) argued that the research programme CURS–UK (Changing Urban and Regional Systems in the UK), supported by the British Economic and Social Research Council, is dominated by the former, i.e. case-study work, and 'exacerbates the tendency to empiricism' (p. 64). Cooke (1987, 77) replied that the programme is seeking to generalize within cases and compare those generalizations, to the extent it becomes possible, with generalizations drawn from within other cases. Then, for those aspects of intra-case generalization which, in concrete terms demand to be understood on a supra-case basis, to be prepared to build on existing theory, concepts, and findings an appropriate, provisional theoretical construct addressing processes which operate on an inter-case basis. (Extract from a section entitled 'Generalization and scale', pp. 76–8)

This looks like an agglomerative method of theory construction, very close to an inductive perspective. Yet, presumably, the specific localities in England chosen for study were so chosen, originally, because of some over-arching deductive criterion—possibly their potentially very different experiences of the common process of restructuring and why.

Perhaps it is easier to demonstrate this scale conflict in the time dimension. In Fig. 4 the general trend XY is upward, but at the microscale AB it is manifestly downward. One could say that AB was an insufficient or unrepresentative sample, but often the sufficiency or appropriateness of the sampling is unknown or cannot be supplied. Suppose point B represents the present, or, at least, the last available data, then mere extrapolation would forecast decline unless some of the general principles causing macroscale growth were already understood.

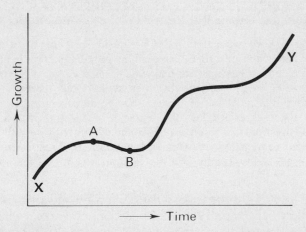

FIG. 4.   *Meso-time-scale trend XY conflicts with micro-time-scale trend AB*

Scale has an important effect when studying spatial segregation of phenomenal subtypes within a total data set. This has a dry-as-dust triteness, but Jones and McEvoy (1978) alleged that a mythical view of race relations in Britain had been given because geographers and others, including an Institute of Race Relations report, had analysed the segregation problem at the wrong scale.

The smaller the scale or observation and the more numerous the territorial subdivisions, the greater the apparent degree of segregation. . . . Where minorities are small in relation to total population, it is essential to use a fine spatial mesh. (p. 163)

A . . . recent study actually takes the GLC boroughs as its data set and, on evidence of location quotients, asserts that segregation of blacks has lessened since 1961. (p. 163)

The author of that 'recent study' defended himself by saying that

it adds an important dimension to the study of ethnic and racial segregation in London to know that the inner boroughs' share of major immigrant groups declined in the decade 1961–71. . . . Jones and McEvoy's suggestion that the 'true dimensions of segregation' occur only at the microscale (1978, p. 164) seems to focus research on only one aspect of a multi-faceted situation. . . . *What is critical, is that the interpretations drawn from such data and analyses are appropriate for the scale employed, and that due recognition is given to the effects that scale will have on the measures used in the analysis.* (Lee, 1978, 366).

This is a particular example of a general problem of scale and segregation measurement recognized by Peach and Smith (1981):

Logically, between two extremes, the size of the areal unit will affect the degree of segregation recorded. When all the population lives in one area there can be no segregation recorded; when each member of the population lives in his or her own subarea, there will be total segregation. . . . as the number of members of a subpopulation and the number of areal subunits approach each other, the degree of segregation recorded will increase. However, this result may be obtained either by very fine areal subdivisions or by using very small subpopulations. The two problems of areal scale and subgroup size derive from a common relationship. (p. 22)

Another example of the scale of data collection affecting results is provided by Chisholm (1960): if proportions of workers who commute are collected at the scale of each dwelling, 'then practically the entire working population would be recorded as commuters, as only a very small proportion work at home' (p. 187); if data are collected on the scale of a whole country, then only workers crossing frontiers are 'commuters'. This scale problem, derived from the irregular lattices on which data may be collected, has been generalized by Haggett, Cliff, and Frey (1977, 348–52) as follows:

In general, the results of statistical tests using area based data are affected by the sizes and shapes of the terrestrial units for which the data are collected. . . . (p. 348) the great variety of grids upon which spatial data are measured implies that, wherever possible, methods must be developed conditional upon the lattice used. (p. 352)

The scale of mesh used to catch and present the spatial data can therefore affect the results, and when this is done with malice aforethought we have gerrymandering of electoral districts (Johnston, 1979a, 172 ff.). A simple example of the contrast between scales of approach is given by the act of shopping when we enter a zone of compromise between what is not too far for us at the microscale and what is not too ubiquitous for the seller's more mesoscale perspective (Fig. 5).

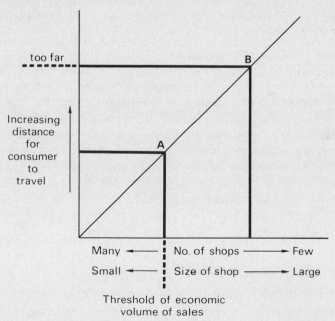

FIG. 5.   *Locational zone of compromise (AB) between a consumer and a shop*

The use of scale in the sense of graphs, horizontal axis compared with vertical axis, is of course very common in geography, but, where proportional relations are being studied over time, there are additional and important implications. Church and Mark (1980) argued that a common approach in contemporary geomorphology was the seeking of 'equilibrium' relations between land-form parameters and supposed governing processes. They pointed out that Bull (1975) had grouped proportional relations under the title of allometric relations. Allometry is 'the study of proportional changes correlated with variation in size of either the total organism or the part under consideration. The variates may be morphological, physiological or chemical' (S.J. Gould, 1966, 629). As far as geomorphology is concerned, Church and Mark go on to distinguish two types of allometry:

Dynamic allometry refers to the study of a single organism (landform) through various stages of growth (at different stages of development). The purpose is to study proportional change with age of the individual example. By comparison, *static allometry* plots data taken from many individuals: the purpose is to study scale-related changes in a population without regard for the course of development of an individual member. (1980, 345)

Dynamic allometry is later used to discriminate between two major models of geomorphology: 'We take 'dynamic equilibrium' to connote a transformation in landscape that yields a self-similar result, and 'evolution' to connote a transformation that yields a non-similar result. The first type of change is isometric; the second allometric' (ibid. 381). And the authors conclude: 'Appreciation of scale relations amongst landform units appear to be a critical part of the exercise in view of our inability to observe directly most significant changes in the landscape' (pp. 381–2).

Finally, as far as human geography is concerned, Watson (1978) has pointed out that behaviour varies between individuals acting alone and in groups of varying size (p. 37).

The gamut of scales and degrees of complexity

The coarse threefold scalar division of spatial patterns pales before the full complexity of scales. Table 6 is based on Boulding's (1956) scheme of systems, which is more accessible in the discussion by Cole and King (1968, 498–9). The levels are orders of functional

TABLE 6. The gamut of scales

Level	Title	Examples or brief description
1	Framework	grid, graticule
2	Clockwork	simple dynamic systems, machines, solar system
3	Steady-state system	homeostatic, negative feedback
4	Open system	all living organisms, negentropic
5	Genetic-social level	plants; division of labour among cells
6	Animal level	increasing awareness and inter-communication
7	Human level	self-consciousness and use of symbols
8	Human organizations	incorporate all preceding levels and exemplify Weaver's (1967) system of organized complexity
9	Transcendental system	realm of ultimates and absolutes, religious faith and conviction in dimensions 'above humankind' if believed in

Source: Adapted from Boulding (1956).

arrangement, and the links between phenomena studied at each level may be at different levels of complexity themselves, posing successively more difficult problems, distinguished by Weaver (1967) as:

1. Ability to deal with problems of *simplicity* containing two factors which are directly related to each other;
2. Ability to deal with problems of *disorganized complexity* by means of statistical mechanics (randomness of patterns suggests randomness of forces);
3. Ability to deal with problems of *organized complexity* where the variables are interrelated (randomness of pattern may mask complex interaction of forces).

Human geography certainly has to deal with problems of this third type and progresses as far as level 8 in Boulding's hierarchy of systems, whereas one might think that physical geography stops at level 6. The superficial inference would be that physical geography is easier than human geography, but this would ignore two important points. First, Weaver's three orders of complexity are found in Boulding's systems from type 4 onwards. Second, physical geographers have increasingly entered the applied realm since 1970, which means they encounter the same social and behavioural problems as faced by human geographers. We may note the opinion of the philosopher Vico that human sciences may be easier than natural sciences, because at least we start from some knowledge about ourselves, whereas the 'book of nature' has been compiled in a dimension outside our own internal frame of reference (Mills, 1982, 3).

Agglomeration and scale economies

The above expression is useful because it encapsulates the close links between economies of scale and the geography of concentration (see the title of Bain's classic paper, 1954; and Table 7). The lessons about economies of scale are provided by economics. Many of them are internal to the operations of enterprises. The degree of pressure to achieve economies is linked to a resultant propensity to concentrate in space. Associated with this is what we might call Bain's Law, and it is worth quoting the original: 'the proportion of the total output of its industry which a plant or firm must supply in order to be reasonably efficient will determine the extent to which concentration in that industry is favored by the pursuit of minimized production costs' (i.e. maximum economies of scale) (p. 15). If the minimum economic plant size increases as a result of the effects of the economies of scale, then this will be another factor in the tendency to spatial concentration. There are of course external economies which link the enterprise to its operational environment, and as there are increases in scale of agglomeration so the opportunity for this type of economy will also increase. There are also economies of scale in the demand function exemplified by consumers' propensity to engage in multi-purpose shopping trips, especially if car-borne (see 'agglomeration effect' in Fotheringhame, 1982, 551, and Fig. 2, 552).

Economies of scale are reached at a point in an upward curve before diseconomies set in, where some exogenous factor overrides the increasing efficiency of internal dynamics of the

TABLE 7. Economies of scale (mainly internal to the operations of enterprises)

1. *Indivisibilities*. Certain component processes which lead to greater economic efficiency cannot be afforded until operations have reached a certain level of output. *Examples*: Provision of R. & D. unit, employment of salesman leading to a sales force; in transport, indivisibilities of investment both in unit of transport and route provision (often result in step-like investments which have to be matched to continuities in upward curve of demand).

2. *Equipment-size economies*. Costs do no increase in direct proportion to capacity. *Examples*: Storage areas, tanks, do not cost twice as much if volume is doubled; growth in average size of cargo-carrying vessels since the Second World War eventually being checked by need to maintain delivery schedules and avoidance of overriding capacity of terminals and depth of sea approaches.

3. *Component process economies*. The greater the total output the greater the output of supporting component processes which then achieve their own economies of scale. *Example*: Car component works (bodies, wheels).

4. *Organizationl economies*. Increases in production result in ability to reap economies of flow-line techniques in more sequences of the operation, to avoid diseconomies of batch production wind-up and wind-down. *Example*: Any industry which has been able to standardize its product range.

5. *Resource economies*. Increase in production machines of similar type does not require a proportional increase in spares holdings. *Example*: If the likelihood of a spares replacement is 1 : 100, a second spare need not be held until the 101st unit is in service.

Source: Bain (1954)

enterprise: for settlements, internal congestion; for industries, managerial inefficiencies or product-range inflexibilities; in transport, bigger carriers may mean longer intervals between deliveries and less flexibility of choice of routes and terminals. There are constantly changing features of production and marketing that alter economies of scale via new thresholds, and this includes any tendencies to product differentiation (fifteen versions of the Ford Sierra) for more discriminating customers.

On the technical side, computerized numerical control (CNC) and computer-aided design and manufacture (CAD/CAM) have given some respite to smaller producers such as Jaguar in the car industry. More dramatically, on the management side, familiar Western ideas that the secret of profitable production lies in the single-minded pursuit of maximum volumes and line-speeds have been turned upside down by the revolutionary Japanese 'just-in-time' system of production which unearths and diminishes formerly hidden sources of diseconomies. . . . Last . . . small plants make for less-adversarial industrial relations. (Sayer, 1985a, 11, in a section entitled 'Economies of scale')

Such counter-tendencies prevent the rise of one megalopolis in every political unit, but the forces in Table 7, along with others, are powerful and continuous promoters of concentration.

Time-scale, mainly via thresholds and corners

A general discussion of scale without reference to time-scale would be manifestly incomplete, though we cannot afford here a long diversion into historical geography and the problems of prediction. But the time dimension is fundamental in geographical studies if they are to be rendered dynamic; and maybe this is one of the reasons for the relative decline of the use of maps in the discipline. Time becomes more interesting when it is used to elucidate 'how' rather than 'when' questions. Stoddart (1966) considered that the idea of change through time had been introduced into geography during the nineteenth century largely through the impact of Darwinism. He explained that the crux of Darwin's theory was the randomness of initial variations and 'a mechanism whereby random variations in plants and animals could be selectively preserved, and by inheritance lead to changes at the species level. In geography, however, Darwinism was interpreted as evolution' (p. 683). '. . . what for Darwin was a process became for Davis and others a history' (p. 688). Many fields of geography became dominated by what can be called the historico-genetic method, which employs the device of 'we know where we are because we know how we got here'. The crucial word here is 'how'. If this explanation is taken to be merely the placing of events in a time sequence, then this is less useful than using time to unfold the workings of processes. The word 'geomorphology' literally indicates a preoccupation with form; Davis brought in the evolution of land-forms through time, whereas some modern geomorphologists would consider 'the study of earth surface processes' a better description of their objective. Nevertheless, a continuing interest in chronology abides because of the length of the time-scales needed to produce much of the earth's surface (K. J. Gregory, 1985, ch. 4).

Figs. 6a and b are inspired by diagrams in a summing-up paper by Schumm (1979). The first is an extract from a modified concept of Davis's geomorphic cycle. A smooth decline in altitude at the macroscale is shown to have more irregularities as the time-scale is decreased. This is analogous to the fact that crises are two a penny in daily newspapers but much rarer in the history of centuries. The nearer we are to an event the more irregular it appears because we naturally tend to study it at the microscale. In 1869 Tolstoy was on to this contrast in the second part of his epilogue to *War and Peace*. In human geography E. Jones (1956) argues that cause and effect in human geography can be more readily observed at the macrolevel of spatial perspective. Tolstoy had observed that deterministic explanations of humanly motivated events are easier to make as the time after the event increases—a contrast resulting from different temporal perspectives similar to the contrast resulting from different spatial perspectives.

Our scale of judgement as to the greater or lesser degrees of freedom and necessity will . . . depend on the greater or lesser interval between the performance of the action and our appraisal of it. . . . A contemporary event seems to us indubitably the doing of all the men we know concerned in it, but in the case of a more remote event we have had time to observe its inevitable consequences, which prevent our conceiving of anything else as possible. (1957, ii. 1433).

Some trends are slow-moving, some fast: it is hard not to see the first from a more deterministic angle, while fast-occurring events appear more anarchic.

Fig. 6b, also from Schumm (1979, Fig. 2, 490, redrawn), is an example of an intrinsic threshold, a state of a system when sudden changes may occur due to slow processes within. A simple example is the state of the camel before the last straw is added. More complex results of the combination of processes are studied by catastrophe theory, which, oddly enough, was first used in historical geography, a subfield of the discipline not usually in the van of the adoption of mathematical techniques (Wagstaff, 1978, and see the bibliography in MacLachlan, 1981).

Fig. 6c indicates a 'corner' (X) on an S-curve which relates to

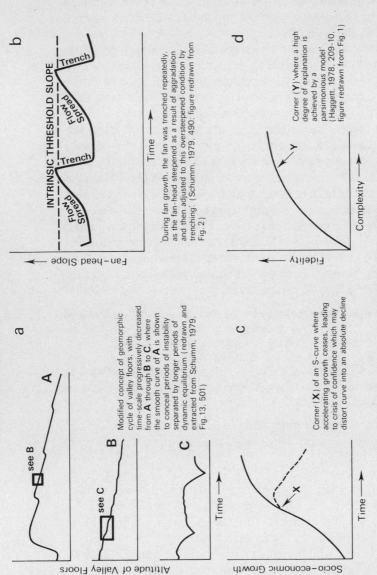

a

Attitude of Valley Floors

A

see B

B

see C

C

Time ——→

Modified concept of geomorphic cycle of valley floors, with time-scale progressively decreased from **A** through **B** to **C**, where the smooth curve of **A** is shown to conceal periods of instability separated by longer periods of dynamic equilibrium (redrawn and extracted from Schumm, 1979, Fig.13, 501)

b

Fan-head Slope

INTRINSIC THRESHOLD SLOPE

Flow Spread   Trench   Flow Spread   Trench

Time ——→

'During fan growth, the fan was trenched repeatedly, as the fan-head steepened as a result of aggradation and then adjusted to this oversteepened condition by trenching' (Schumm, 1979, 490; figure redrawn from Fig. 2)

c

Socio-economic Growth

X

Time ——→

Corner (**X**) of an S-curve where accelerating growth ceases, leading to crisis of confidence which may distort curve into an absolute decline

d

Fidelity

Y

Complexity ——→

Corner (**Y**) 'where a high degree of explanation is achieved by a parsimonious model' (Haggett, 1978, 209-10; figure redrawn from Fig. 1)

FIG. 6.   *Thresholds, corners, and the time-scale*

Note: In connection with Figs. 6a and 6b, K. J. Gregory (1985, 161–3) has pointed out the seminal nature of the paper by Schumm and Lichty (1965), initiating a new attitude to scale as a reconciler of apparently conflicting views in geomorphology between dynamic equilibrium theory and results of investigations into long-term environmental change. 'We believe that distinctions between cause and effect in the molding of landforms depend on the span of time involved and on the size of the geomorphic system under consideration. Indeed as the dimensions of time and space change, cause-effect relationships may be obscured or even reversed, and the system itself may be described differently' (Schumm and Lichty, 1965, 110).

some form of socio-economic growth. Any curve that is concave upward will eventually have to flatten out. But if people get used to increasing growth, a 'corner' which indicates change to declining growth may cause a crisis of confidence about the future, and this may accelerate the new trend, maybe even changing it into absolute decline. This is an example not merely of a self-fulfilling prophecy but also of a more general fact that prediction in the social sciences may in itself be one of the variables that can cause change.

Fig. 6d represents two compared scales of fidelity and complexity (derived from Haggett, 1978). The use of the word 'fidelity' prompts an illustration from the world of domestic hi-fi. Some systems on offer are well to the right of the ideal compromise corner (Y), in areas where great increases in expenditure produce only small increments of sound fidelity that may not even be detectable by fallible human ears. The trade-off between fidelity and complexity might also be generally illustrated by the number of axes in scalar comparison. Two scales give easy visual comparison, but as more axes are used in models so complexity increases until a wish arises for a return somehow to a more simple exposition.

The politics of scale

Scale seems like an innocent field for a geographer wherein it is not too difficult to remain objective in one's choice of approach. But when more than one scalar level is considered, we are soon led to the idea of scalar hierarchy; and Webb (1976, 17) has usefully distinguished 'underimposed hierarchies', such as household–street–neighbourhood–community, from 'imposed hierarchies', such as district–city–county–state, an overtly political hierarchy. This leads us to political geography, where there seems to be a general consensus on three scales of study: international, national, and intranational (see Table 4 and P.J. Taylor, 1982, 21–3 for references; and Short, 1982, 1 and 4–5). If for 'international' we read capitalist world economy, then there has been a suggestion that radical geographers will tend to enter the three-tier system at the extensive level because to do otherwise would be to take the capitalist system as given. Others have denied this link between ideology and choice of level: you can break into the system of causation 'below the level of the system as a whole' and remain radical (Massey and Meegan, 1985, 6–7). Taylor (1982) certainly finds that for his materialist approach the determining scale is the world economy 'with the constraints imposed by the needs for maintaining capital accumulation' (p. 23). He rather conveniently contrasts this with Philbrick's (1957) principles of areal functional organization, which use the aggregative approach based on units which are an 'outgrowth of human creative choice' (p. 300). Taylor accordingly calls Philbrick's nested hierarchy of functional organization an 'extreme liberal theory of geographical scales' (p. 23).

So choice of the determining scale of approach may result from choice of a political perspective. Taylor (1982, 24–5) goes even further and suggests that the *number* of scales in the world-view may have political overtones. Relying on the work of Wallerstein (1975), he argues that a three-system format is essentially self-stabilizing and therefore supportive of the dominantly capitalist world economy, whereas opponents of this structure 'attempt to polarize the situation into just two sides'. We are here close to the argument that to adopt a 'value-free' approach is nevertheless to adopt an ideology: the ideology of the so-called 'value-free'; 'so-called' because by default it is supportive of the status quo, and the global status quo is capitalist-liberal. Work in political geography has led P.J. Taylor (1987) to what he calls 'the paradox of geographical scale'.

This is the notion that whereas the social and economic forces that are continually changing our world are manifestly global in scope, the vast majority of political actions take place at the altogether smaller scale which is the nation-state: . . . the geographical paradox becomes, in world-systems analysis, a surface manifestation of a basic antinom in the capitalist world economy: classes *für sich* organised at the state scale and classes *an sich* defined globally (Wallerstein, 1984, 36). Since classes express their consciousness at a geographical scale that does not reflect their objective economic roles there will be a general tendency for political behaviour to be contradictory. I have tried to capture this interpretation of politics in my political geography by devising a framework based on geographical scale. In this argument the state represents the scale of ideology, separating the scale of reality (the world system) from the scale of experience (the local scale which covers our day-to-day activities). (p. 287)

While agreeing with Taylor's threefold scalar division, N. Smith (1984, 135–6 and 176 *n.*) believes 'there is a more directly "materialist" framework' for understanding these distinct spatial scales under capitalism. While the discrete nature of the scales predates capitalism,

> The scales themselves are not fixed but develop (growing pangs and all) within the development of capital itself. And they are not impervious; the urban and national scales are products of world capital and continue to be shaped by it. But the necessity of discrete scales and of their internal differentiation *is* fixed. This provides the last element in the theory of uneven development. (p. 147)

In practical terms it is hardly possible for every study to start with the world-view and disaggregate down to the scalar level appropriate for the objects of study. But it is as well to remember that one's political views, implicit or explicit, do affect the perspective; the objects of study are inclined to look different depending on whether you implicitly support or are actively opposed to the higher tiers in which they are set.

Scale compromises?

A recent comparison has been between extensive and intensive research (Sayer and Morgan, 1985); but these two terms are not synonymous respectively with macroscale and microscale. It is as though the scalar terms measure the form of the approach, whereas the extensive–intensive polarity describes the functional framework adopted where each asks different sorts of question and defines its objectives differently. In practice, extensive research does deal with macroscale taxonomic groups and their similar or dissimilar relationships, whereas intensive research in causal groupings studies individual agents in their causal contexts. But because of their different explanatory frameworks it is not easy to envisage a synthesis of extensive and intensive research.

If we compare macroscale and microscale approaches, we may wonder if they have complementary advantages and about the possibility of their integration. Watson (1978) considered this problem in great detail, and from her argument we can construct the list of advantages and disadvantages in Table 8. Here we concentrate on her conclusion which focused on three strategies

TABLE 8.   Relative advantages and disadvantages of macroscale and microscale approaches

MACROSCALE	MICROSCALE
Advantages	**Advantages**
When exploring possible regularities in *terrae incognitae* more economical to begin at macroscale	Pays attention to crucial microvariables
Uncovers macroscale regularities, pays attention to socio-structural context	Allocation of resources often determined by a few powerful individuals, politicians, interest groups
Where behavioural repetitious or indeterminate	Action preferences, beliefs, values, attitudes, decisions best studied at microscale
Where environmental constraints make behaviour predictable	Which societal goals and decisions about them often best studied at microlevel
Relationships downwards to subsystems which may be not much related to each other	
Emergent variables become visible only at certain levels of aggregation	
Prediction better at macroscale	
Easier to disaggregate macroscale than aggregate microscale	
Both statistical inference and logical inference[a]	
Disadvantages	**Disadvantages**
May ignore crucial microvariables	Lacks advantages inherent at macroscale
Cannot cope with equifinality	Makes assumptions about the macro-environment impinging on individuals and then infers aggregate behaviour from conclusions
Provides prediction without explanation	
Stationarity assumptions less tenable as time-scale increases	Logical inference only (inference from case studies)[a]
Probability functions often arbitrarily chosen	

TABLE 8. Continued

MACROSCALE	MICROSCALE
In society no given wholes; in reality data may require considerable modification before aggregate techniques of analysis can be employed	

[a] This distinction has been added from Mitchell (1983). Logical inference constitutes 'the process by which the analyst draws conclusions about the essential linkage between two or more characteristics in terms of some systematic explanatory schema—some set of theoretical propositions' (p. 200; see also S. Smith, 1984, 359).

Source: Based on Watson (1978).

for dealing with the two scales: aggregation, disaggregation, segregation. She described the satisfaction in trying to show

how the behavior of individuals combines to produce aggregate behavior. Disaggregation is attractive in situations where the macro theory is better developed than its counterpart, or where the mathematics of aggregation appear intractable. Segregation is a strategy for linkage which preserves the two levels intact and tries to ensure consistency between them. It leaves the researcher free to confine attention to one level, while ensuring that the model is consistent with, and benefits from developments at the other level. (p. 47)

This advice seems sound, though segregation hardly describes the recommended strategy, for we may lose advantages if we 'confine attention to one level'. First, we must choose the level that is thought a priori best for the problem data; and perhaps our choice is guided by past experience of that sort of data, or by whether our principal objective is synthesis or analysis, or we may have a political view that gives the 'determining' scale. Whatever a priori choice we make as our scale or level to confront the problem, it may not be as error-free as we suppose; so we have to keep glancing continually at the contrasting scale of operations, to see if we can minimize the disadvantages of our scale of approach and to reap some of the opposing scale's advantages (cf. Ambrose and Williams, 1981, 13).

One way of doing this is to modulate to another scale of approach during the course of the study (Fig. 7). The following is from an account of the modulation procedure in music, with

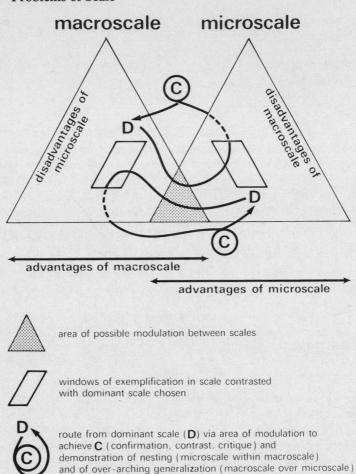

FIG. 7. *Scale modulation*

insertions to show the relevance to a scalar strategy: 'Modulation may be familiarly defined as a method of key- [scale-] change without pain. The adoption of the new set of tonal [scalar] conditions is softened. For instance, the "take-off" into the new key [scale] may be judiciously managed from a chord [case] that is common both to it and the old key [scale]' (Scholes, 1942, 589). This is not to recommend eclecticism, or synthesis of scales, or a dialectic of scales, or any process of oscillation that may go under such names as reflexivity or recursiveness. We should choose our dominant scale of approach and stick to it, modulating to the other scale only to offset the disadvantages of our approach, offering perhaps a vivid illustration and a running critique of some of our work, and putting a tension into it. We should be able to justify the strategy by the usefulness of the results achieved. In the course of our study we may find ourselves in error; the criticisms generated by the other scale of approach may overwhelm our a priori strategy. A new start is then necessary with another dominant scale of approach. In our submitted presentation the dominant scale will be apparent. We will have made a comparison and a judgement. Such a strategy may help us when we come to the even more serious task of confronting various philosophies that are or have been current in geography.

3

Alternatives to Scientific Method in Geography

Before the parade

Assume that Chapter 1 was correct and that 1966 was the year in which it was finally demonstrated that all geographical data, even 'unique' locations, are amenable to scientific method, as an expression of modern positivism (see Table 9). The question immediately arises as to how long this methodology remained overwhelmingly dominant in the discipline. This is not an easy question to answer because in the 1960s and 1970s many geographers believed that the fundamental change in the discipline was an increased reliance on quantitative techniques. Even if it was felt that number-crunchers were missing some essential qualities in the data, to voice criticisms might have been taken as an expression of numerical incompetence. Of course, physical geographers were less inclined to find fault with the hypothetico-deductive method, particularly if they were largely concerned with the purer forms of physical geography. But, subsequently, they encountered many of the -isms to be mentioned in this and the following chapters as they moved into the applied realm. A landmark publication in 1969 by Harvey was definitely scientific in aim with its last words: 'By our theories you shall know us' (1969*a*, 486). But during the 1970s, doubts began to surface so that, by the end of the decade, Hay (1979) felt it necessary to respond overtly to criticisms of positivism. Other -isms had arisen in geography and were fighting for their place in the sun, often attempting to trim the scientific shadow. Perhaps the most visible of the earliest attacks on the whole conception of science in geography was Zelinsky's (1975) presidential address to the Association of American Geographers:

May I suggest that the fundamental reason for our [scientific] demigod's dilemma is simple and highly disconcerting. In grappling with the

TABLE 9.   Johnston's four philosophical categories relevant to human geography[a]

Political stance[b]	Approach	Ontology[c]	Epistemology[d]	Methodology
Status quo	Empiricist	Experienced things exist as fact	Know via experience	Presentation of experienced facts
Status quo	Positivist	Agreed verifiable evidence	Know via experience based on verifiable[e] evidence	Scientific methodology
Liberal reformers	Humanist	What exists is that which people perceive to exist	Knowledge obtained subjectively in a world of meanings created by individuals	Investigation of individual worlds = emphasizes individuality and subjectivity rather than replicability
Radicals	Structuralist	What really exists (i.e. forces or structures creating the world) cannot be observed directly but only through thought[f]	World of appearances does not necessarily reveal world of mechanisms (which causes world of appearances)	Construction of theories which can account for what is observed but which cannot be tested because direct evidence of their existence is not available

[a] Johnston (1983b, 5, and chapters 2, 3, 4); but beware dangers of this tabular compression.
[b] Added to Johnston's scheme and often relevant in those societies which allow free political expression as the norm.
[c] What is believed to exist and how we can come to know about it.
[d] What is supposed to exist, what one believes, grounds of knowledge, what we can know. For relationships between ontology, epistemology, and methodology, see Harrison and Livingstone (1980, Fig. 1, 27), 'The presuppositional hierarchy'.
[e] Popperian science would substitute the term 'falsifiable' evidence, i.e. 'knowledge' alters under criticism and progressively as new testable theories replace older discarded theories.
[f] 'What I have thought, you must believe' is a possible degradation. (My note.)

complicated, abstracted, fluid, and intersubjective data of the social realm, he has been applying a fatally inappropriate model to the world of human beings, an utterly useless, even damaging way of thinking about it. The model in question is, of course, the outmoded mechanistic natural science model: . . . in brief, [the social scientist] . . . has been asking the wrong questions about the wrong things in the wrong way. (pp. 139–40)

Often those who had espoused the philosophy and methods of the natural sciences with the greatest enthusiasm were among those who later espoused other philosophical standpoints with equal fervour, as a result perhaps of a 'whole-hogger' tendency embedded deep in their personalities. What they left behind they dubbed as 'positivistic science' or 'positivism', although the latter term has a long history as a philosophy dating back to its founder Comte (1798–1857). It is common practice now to take scientific method and modern positivism as synonymous, with the following five major features:

- objectivity (via a value-free methodology),
- hypothetico-deductive method (via theory-laden observation),
- testability,
- replicability,
- predictive ability.

Five major criticisms of this methodology have been:

1. A value-free methodology is itself an ideology, which because of its alleged neutrality supports the societal status quo by default and is therefore more likely to be an ideology right of the political centre in Western societies.
.2 It is disabling to assume that the observer, the 'scientist', can be objectively separate from the data observed.
3. Theory-laden observation views data in isolation from the surrounding 'noise', and the resultant gains of selectivity and canalization of effort may amputate what may be an essential context.
4. The aim of a human geographer is not the explanation of phenomena but the understanding of the actions of fellow human beings.
5. To assume a reality independent of the observer is illogical because the observer is always part of the 'real world' being observed (see 2 above).

A positivist geographer might reply as follows:

1. The assumption of a value-free methodology is a thought experiment (via a hypothesis) of the enabling 'as if' kind, which can get results by acceptable methods of testing. If it can be shown that the methodology in fact vitiates the results, that is a criticism which is sustainable or not, and all work is subject to criticism.
2. If the scientist assumes that he is separate from the data, that is another of the assumptions embedded in the hypothesis to be tested.
3. Because everything is connected to everything else, all methodologies have to put boundaries between themselves and an endless context.
4. The scientific method does not preclude the insertion of illuminative human detail exemplifying important issues.
5. It is true that the eye tends to see what it seeks, but in theory-laden observation it should, nevertheless, be possible for the data to overturn the initial hypothesis.

Even those who might accept the above defence of positivistic scientific method as reasonable would admit that there is obviously a good deal of room for alternative philosophies and methodologies. For the moment I am going to place all these under the two headings of humanism and structuralism, however uncomfortable some of them may appear later.

Macroclassifications of philosophies: a triad, a dichotomy, and some polarities

In their exploration of social geography, Jackson and Smith (1984, 4–5) found it useful to describe a triad: 'various perspectives exploited by social geographers can all be illustrated with recourse to the philosophical triad (of positivism, humanism, and structuralism) which we take as our major theme'. Fig. 8 actualizes this 'imperfectly connected triangle' (ibid.); and Table 9 summarizes a somewhat similar classification used by Johnston (1983*b*, 5; and see his chapter layout, chs. 2–4). In proposing such a triad Jackson and Smith warn against attempts to synthesize one scheme by some form of eclecticism or conflation of fundamentally different philosophies, and they quote the now well-known warning of Eyles and Lee (1982, 117):

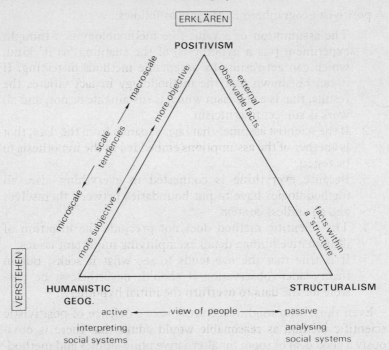

FIG. 8.   *The Jackson–Smith triad*

Note: Such a triad and some of the annotations are suggested in the text of Jackson and Smith (1984, 4–5), but they are not responsible for this particular 'actualization', nor for all the annotations and their positioning. Scale tendencies have been inserted to suggest that some of the contrasts between scales of approach are similar to contrasts between epistemologies.

There appears to be an assumption . . . that these 'approaches' are in some way alternatives; that we can select this concept from one approach, that method from another, and the perspective from yet another. Unfortunately this desire for eclecticism rests on the false assumption that these approaches are above all techniques for analysis rather than epistemologies. In other words, there is a failure to recognize that the three 'approaches' are in fact different philosophical systems, all demanding different modes of validation.

It is also worth recalling an earlier diagram from Chorley (1976) where there is a dichotomous classification of positivism and 'behaviouralism' and which has the added advantage of a simple time dimension (Fig. 9). In the 1960s, movement over time would

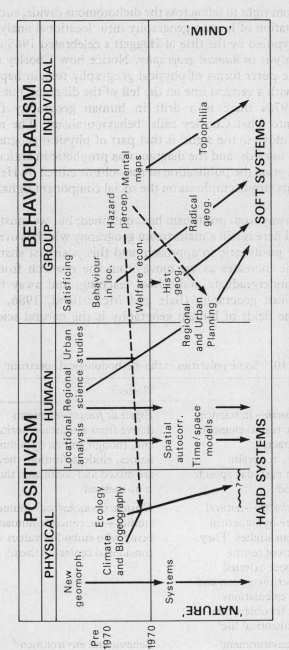

FIG. 9. *Geographical studies arranged according to relative 'hardness' or 'softness'.* This is an extract redrawn with permission from a diagram without a legend in Chorley (1976, 31)

have been from right to left across the dichotomous divide, such as the transmutation of human geography into 'locational analysis', beautifully typified by the title of Haggett's celebrated 1965 text, *Location analysis in human geography*. Notice how Chorley suggests that the purer forms of physical geography remain happily positivistic with a vertical line on the left of the diagram, but that since the 1970s there is a drift in human geography from positivism into what Chorley calls 'behaviouralism'. The most spectacular 'move to the right' is that part of physical geography dealing with hazards, and the diagram was prophetic in anticipating by seven years the publication of the volume edited by Hewitt (1983) with its strong emphasis on the social component in hazard studies.

The drift away from positivism has continued; but we must not exaggerate. There is still a mainstream geography which is overtly or implicitly positivistic in approach, and this positivist share of the discipline increases as we move from the research frontier down into undergraduate and school teaching, and away from Anglo-American geography (Bale and McPartland, 1986, 63). Only in some fields of human geography is the natural science

TABLE 10.   Some polarities in the methodological spectrum[a]

'Harder'[b]	'Softer'[b]
Esprit de géometrie—in science (the geometric realm) elements are clear, abstract, unchangeable, in a realm removed from everyday speech: there is method[c]	*Esprit de finesse*—elements derive from a common heritage and, though known by common names, elude definition: they are mixed and confusing: there is no method[c]
Naturwissenschaften—natural sciences—more hypothetical than the human studies. They have the means of testing hypotheses about external nature by experimentation and mathematical calculations which are not feasible when dealing with historical life[d]	*Geisteswissenschaften*[d]—human studies—the concrete human being who embodies values is considered central to them[e]
Phenomenal environment[f]	Behavioural environment[f]
Erklären[g]—explanation in the	Verstehen[h]—understanding in

TABLE 10. Continued

'Harder'[b]	'Softer'[b]
natural sciences	the human sciences
Environmentalism[i]—world of objects[j]	Existentialism[k]—world of beings. Life world[l]
Categorical paradigm[m]	Dialectical paradigm[m]
Empirical-analytic form of knowledge[n]	Historical-hermeneutic form of knowledge[n]
Taken-for-granted world[o]	Emerging world[o]
Convergent thinking[p]	Divergent thinking[p]
Deductive	Inductive
Theoretical language ←——→ Observation language	

Reflexive relation[q]

Expected ←——————→ Experienced

Diagonalism[r]

[a] Many of these polarities owe much to Gregory (1978) and page numbers of references to this monograph appear in square brackets below.

[b] See Fig. 9. The comparative adjectives are used in the following senses: 'harder' refers to the positive end of the methodological spectrum; 'softer' to behavioural-type studies.

[c] A polarity from Pascal. The notes are a free translation derived from a 'compressed paraphrase' by Barzun (1974, 91), itself and all-out attack on positivism in historical studies. A convenient source for Pascal's *Pensées* is the edition by Sellier (1976, 345).

[d] Makkreel (1975, 60) [59].

[e] Makkreel (1975, 42) [59]. The notes for this polar pair are based on Dilthey (1924, republished 1957, 143–4).

[f] Kirk (1951, 1963) [59]; cf. Popper's three worlds, see Bird (1977b, 104; 1985).

[g] 'We explain [*erklären*] nature, but we understand [*verstehen*] psychic life,' Dilthey (1957, 144) [60, 133].

[h] Dilthey (1957); see also Harvey (1969a, 56) and Rose (1981) [59–60, 132–3].

[i] Tuan (1971, 182) [60].

[j] Cf. Popper's World 1, see Bird (1977b, 104; 1985, 404).

[k] Tuan (1971) [60].

[l] Buttimer (1976).

[m] Olsson (1975, 60–2) [65].

[n] [70, Table 1].

[o] Schutz (1967, 35–6); see also Schutz and Luckmann (1974, 3 ff. and 21)

[p] Hudson (1966, ch. 3). ('... geography ... attract[s] convergers and divergers in roughly equal proportions', p. 42.)

[q] [Fig. 2, 58, 76].

[r] Bird (1978, Fig. 2, 137).

Source: This table first appeared in Bird (1979, 119).

model held as a minority view. If I overstate the case, let us just say that positivism still has a strong hold within the discipline, so it is not too far-fetched to write of 'positivism and the alternatives' and to remember the long-standing polarity of *erklären* (explanatory-positivistic) methods of investigation and *verstehen* (sympathetic understanding) approaches (see Table 10). But it must be admitted that triads, dichotomies, and polarities, however useful as dramatic summary headlines, miss a great deal of the rich variety of various epistemologies. Johnston (1983*b*) has a book-length treatment on the three philosophies of positivism, humanism, and structuralism, which he terms 'approaches' 'because each embraces a variety of related viewpoints' (p. vi, with the barest summary given here in Table 9).

Moving away from positivism in geography

It is now possible to reorient the Jackson–Smith triad by viewing the approaches of humanistic and structural geographies as alternatives to scientific method in geography or as promoting alternatives to the status quo in Western societies. Many geographers who have worked under these frames of reference would no doubt resent being classified under headings that are seen as merely reactions to something else. After all, it would be possible to construct tables with the headings, 'Alternatives to humanistic geography', 'Alternatives to structural geography', in which positivism and the scientific method were given a similar apparently subordinate position. The actual scheme which is offered here (Fig. 10) has its justification in the dominant position that positivism has occupied in the discipline and is put forward merely as a first headline-making step. The next two chapters will give room for more detailed exposition of these arbitrarily named 'alternatives'. There is another danger in offering such a scheme as Fig. 10: it looks as though the alternatives are being ranged in some sort of continuum or at least in the same two dimensions. Actually, some of the frames of reference are in different worlds, deriving from fundamentally different ways of looking at 'reality'. This does not mean that we cannot cross over from one frame of reference to another.

Many a geographer who now embraces an epistemology which sits under the broad headings of humanism or structuralism once

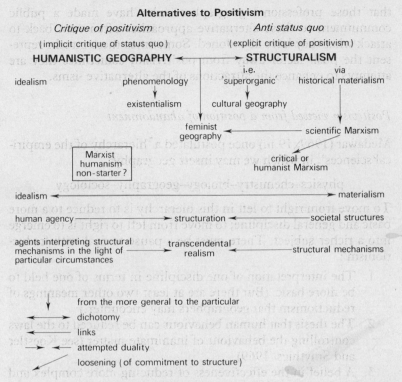

FIG. 10. *Alternatives to positivism in geography*

published squarely within the realm of positivism. Many began their career with a first degree based on a mainly positivistic undergraduate curriculum. They have changed their epistemological stance: either as a natural evolution of their studies; or because they became dissatisfied with what they considered an inhuman or insensitive view of the world; or because they became disillusioned with the society in which they lived and wished to adopt a view which advocated societal change; or a combination of any of the above. Perhaps some were radical in outlook from an early age, but only felt strong enough to 'come out' when they had emerged from what they would call positivistic brainwashing. Students who have not previously encountered philosophical debate often quickly become interested in these alternative approaches on realizing that there are worlds of thought that are not dominated by so-called factual evidence and precise measurements. They find that those professional geographers who have made a public commitment to these alternative approaches often turn back to attack what they have abandoned. Sometimes these attacks represent the push factor away from positivism; sometimes they are attempts to enhance the attractions of the alternative -isms.

Positivism viewed from a position of abandonment

Medawar (1969, 19 n.) once postulated a 'hierarchy of the empirical sciences', in which we may insert geography:

physics–chemistry–biology–geography–sociology

To move from right to left in this hierarchy is to reduce to a more basic and general discipline; to move from left to right is to emerge into a richer subject. There must be a pause to deal with 'reductionism':

1. The interpretation of one discipline in terms of one held to be more basic. (But there are at least two other meanings of reductionism that geographers may encounter.)
2. The thesis that human behaviour can be reduced to the laws controlling the behaviour of inanimate matter (see Koestler and Smythies, 1969).
3. A belief in the effectiveness of reducing more complex and macroscale phenomena to less complex microscale phenomena, notably as the preference for study of the individual over that of groups or of societal structure.

With reference to this last meaning, the opposite belief in the superior effectiveness of the study of structures is often dubbed 'reification' by critics. In the present context 'reduction' is the antonym of 'emergence' as in the following:

There is a sense in which the social sciences comprehend biology and make use of biological notions, and in which sociology is empirically and conceptually the richer subject. The same could be said of the biological sciences *vis-à-vis* physics and chemistry [emergence]. Yet there is also a sense in which physics comprehends biology, and biology in turn the social sciences, as the more general sciences comprehend the more particular [reduction]. (Medawar, 1969, 16 n.)

This seems quite broad-minded but notice that the emergence and reduction take place in a linked hierarchy. This view of the unity of method across disciplines is repeated within a ten-point self-assessment positivism test (Hill, 1981). The idea is that the more inclined you are towards positivism, the higher you score. I have administered this test to groups of student geographers, generally confirming the following hypothesis: that the more physically oriented geographers score higher on this test than those more inclined towards human geography. But, in justifying a positivistic answer on point 4. of the test, Hill suggests that we shall eventually be able 'to connect the findings of all the empirical sciences in a single deductively unified system. . . . it is literally unjustifiable nonsense to insist that the natural and human sciences deal with different kinds of reality' (pp. 55–6). So it is 'unjustifiable nonsense' to object to a linked system of sciences. But many human geographers do not want to conform to such a hierarchy of the sciences; some want to enter and use entirely different perspectives and methods of study; others go further and want not to explain the world but to change it. Some of these 'wants' derive from the logic of specific, non-positivist epistemologies (see Table 9); others derive from acts of faith, with *post hoc* rationalization. These are part of currents in the social sciences which have given rise to all the alternatives to positivism and scientific methodology that are paraded in the following two chapters. In advancing these other -isms, the advocates have often turned to attack positivism, and six examples are given below, from a philosopher of science, a political scientist, and four geographers. You will see that strong words like 'unjustifiable nonsense' can be matched by some of these opponents of positivism.

I begin with the philosopher Paul Feyerabend (1975), and his very title is an attack on defenders of scientific method or positivists: *Against method: outline of an anarchistic theory of knowledge*. By 'anarchism' Feyerabend means the suspicion of a philosophy that is ruled by some abstract goal such as the scientists' 'search for truth' (though Popper has cleverly softened this by using the wonderfully appropriate word 'verisimilitude'); that 'theoretical anarchism is more humanitarian than its law-and-order alternatives' (p. 11); and here is a flavour of the attack on positivism.

. . . late 20th-century science has given up all philosophical pretensions and has become a powerful *business* that shapes the mentality of its practitioners. Good payment, good standing with the boss and the colleagues in their 'unit' are the chief aims of these human ants who excel in the solution of tiny problems but who cannot make sense of anything transcending their domain of competence. Humanitarian considerations are at a minimum . . .

Scientific laws can be revised, they often turn out to be not just locally incorrect but entirely false, making assertions about entities that never existed. There are revolutions that leave no stone unturned, no principle unchallenged. Unpleasant in appearance, untrustworthy in its results, science has ceased to be an ally of the anarchist and has become a problem. (pp. 188–9)

The problem resides in science considering itself as an orthodoxy which dubs as unscientific those that do not subscribe to its general frame of reference; and, moreover, to be unscientific is to be an unperson not worthy of being debated with. Whereas in the Popperian version of the scientific method there is a libertarian, open-ended competition between scientific theories, even there, if the general hypothetico-deductive method is not followed, the procedure is called metaphysical, pre-scientific, or non-scientific. Science is ruthless in permitting no competitors, and this has political implications, as a political philosopher has observed: 'Scientific knowledge, positivistically conceived, is inherently repressive, and contributes to the maintenance of a form of society in which science is one of the resources employed for the domination of one class by another, and in which the possibilities of radical transformation towards a more rational society are blocked and concealed' (Keat, 1981, 2).

This is one of the quotations assembled by Mercer (1985), who, after a previous essay attacking technocratic geography (1984), here turned to look at the way physical geographers have 'read the book of nature' following the general natural science model:

Put simply, the principal—and doubtless, to some provocative— argument of this paper is that mainstream professional geography has become locked into a particular kind of conservative, scientific frame of reference—or straitjacket (Martin, 1981)—which inevitably leads to a blinkered, cornucopian view of the 'environmental crisis', shores up a technocratic approach to problem solving in the 'real world' and precludes certain fundamental research problems from being asked. This is by no means the first time that such a criticism has been made. In 1977, for example, in what has since become something of a 'classic' review, Waddell wrote a devastating critique of the overt scientism and naivety characteristic of most of the extensively funded research associated with

the Gilbert White school of 'natural hazard' studies. Waddell lays bare
the transparent ideological bias of the investigations and highlights the
affluent, western attitudes towards 'nature', 'hazards' and development
that are uncritically assumed throughout. A similar viewpoint is
expressed in Pepper's (1984) excellent new book. (pp. 9–10)

In 1975 Olsson, who had published extensively within the posit
-ivist frame of reference, publicly repudiated it: 'I lost interest in
location theory' (1975, p. vii; see also Olsson, 1980). He had experienced the 'tension between the pragmatist's quest for certainty
and the rebel's recognition of ambiguity' (p. ix). The following
are just three brief flashes from a book-length treatment describing this tension.

. . . the intellectual race course. . . .

Those who race are different from ordinary people. To hide their
feelings, they dress in strange outfits. Their helmets are designed to hold
their heads in place. Their goggles are cut to ignore what can not be
counted. They have left their hearts at home, for otherwise they can not
be objective. They feed on a diet of certainty and they get upset by
ambiguity. (p. 472)

Through planning based on descriptive models, we consequently run
the risk of imposing on reality a strictness which it neither has nor ought
to have. If we are so protective of our image that we refuse to recognize
this hallmark of positivistic methodology, then we shall be left with
a society which mirrors the techniques by which we measure it.
(pp. 495–6)

And earlier:

The conclusion is that scientific explanation is marred with difficulties,
most of which are related to errors of measurement' (p. 427).

And one example he gives is that of the rise and fall of the gravity
regression model, which started slowly and thoughtfully, then
screamed from overextension, but still roared ahead (p. 475).

Seamon (1984) has attacked positivistic research for destroying
the very 'thing' that it is studying. The researcher organizes the
phenomenon or problem in question into 'imposed parts and relationships which most conveniently fit the research strategy'
(p. 217). There is a tendency to pay attention only to those parts of
the 'thing' that can be measured. An example of this was the
method of assessing the quality of research in departments of
British universities during 1986. Not only was great weight given

to the measurable feature of research funding achieved, but there was evidence that this was counted twice (D.M. Smith, 1986, 248)—an example where 'the need for quantification transcends the validity of the indicator adopted' (ibid.). Positivists might reply that at least what they do is based on measurable evidence, but Seamon has a sharp word for that defence, and then the words grow even sharper:

The thing may suffer even greater humiliation: it can only be what the researcher has the means to measure. Woe to some dimension of the thing which cannot fit into the researcher's framework or which cannot be identified empirically! Such aspects of the thing must be discounted . . .

The manipulative, explanatory, predictive style of positivist science closely parallels the masculine, centralized, materialist power structures which not only subjugate minorities and even entire national populations but also manipulate and exploit the natural environment, bringing on ecological damage and collapse. (1984, 217)

By this point the positivist is either reeling or in the process of building his epistemological defences. Non-positivists and non-physical geographers may be nodding calmly in approval, but even they are not spared in an attack by Eliot Hurst (1985), who states that the whole discipline has been on the wrong route. Human geographers have been offering causes for spatial patterns which are secondary (*causa causata*). By confusing cause and effect the true causes (*vera causa*) have been overlooked. These are derived from the dominant mode of production: 'The striving to create new opportunities for capital accumulation; a struggle to realize surplus value; the crystalization of patterns of over-development and underdevelopment; the need to create global markets . . . (p. 82). Eliot Hurst would build a new focus grounded 'in the praxis of Marxism', in which all geography, as at present and potentially practised, would be swept away.

There are obviously some strong philosophical and method-ological winds a-blowing, and I hope you now feel up to reviewing the parade of -isms in detail. If so, by all means skip to the next chapter. But you are invited to pause awhile to consider a gentle comparison between two contrasted frames of reference. This comparison is far removed from politicized forms of recent debates; and yet gives a hint that there is more than one way of going about matters. If you are growing impatient and are saying to yourself that a choice has to be made sooner or later, all who choose, whatever they in fact do choose, might agree that an informed choice is better than a mere first alighting. And I believe that, even as we pause, a little more information is being painlessly provided.

A gentle pause: erklären compared with verstehen

I now try to demonstrate an example of the contrast between *erklären* (explanation based on cause and effect, appealing to the intellect) and *verstehen* (impressionistic, sympathetic understanding by means of ambiguous stimulation of the reader's imagination). I shall use two types of regional description of the area shown on Fig. 11: first, as a geographer might expound a view to a field party on a hilltop; secondly, I quote a passage from E. M. Forster's novel *Howard's End*, where he describes the view from an identifiable spot height.

The geographer would, I hope, be concerned to relate the obliquely viewed disc of the scene to the wider regional context. He would find himself trying to demonstrate cause and effect; it is interesting and informative to attempt an explanation of the origin and development of land-forms. But in the world of *erklären* one should beware of what I have called 'mild physical determinism' (Bird, 1983a, 58–60)—and I must record once being guilty of the following:

An important task in geography is the elucidation of the physiographic stage upon which man has erected his own scenery. The nature of the physical environment does not determine the use which man shall make of it; *its influence* varies from place to place and at different times in the same place. It is difficult to generalize about its role except to say *that the influence of the physical setting* is likely to be strongest in the earliest days of its use, because the smaller the number of people and the less developed their technology, the more they are likely to be *influenced* by the natural advantages and disadvantages of their immediate environment. (Bird, 1957, 15; italics added)

The words in italics suggest an active influential role for the physical environment, which is wrong. Later, I was more circumspect and avoided mild physical determinism by the following: 'man's actions are not determined by the environment but by what he thinks the environment, physical and social, determines him to do' (Bird, 1977a, 6). Notice that my 'geographical' description below assumes understanding of the technical use of such terms as 'discordant', 'superimposed', 'antecedence'. And during the whole of the passage we have to remember that the human occupants have moulded their patterns over the centuries as they saw the opportunities on offer from the physical environment and the human legacies from the past.

Geographers have long paid tribute to the insights available from regional novelists. In the second passage we see such an artist at work. He is highly selective, yet we seem to see more from the viewpoint than actually can be seen. We are now in the world of *verstehen*; and the 'real' scene seems to stand for much of what cannot be seen. Moreover, the rhythm of the words induces an hypnotic effect on the reader, like overhearing a semi-magic incantation. Forster was himself conscious of the fact that he was using words to create an atmosphere which

resides not in any particular word, but in the order in which the words are arranged . . . It is the power that words have to raise our emotions or quicken the blood. It is also something else, and to define that other thing would be to explain the secret of the universe. This 'something else' in words is undefinable. It is their power to create not only atmosphere, but a world, which, while it lasts, seems more real and solid than this daily existence of pickpockets and trams. (Forster, 1951, 80–1)

The ideas are not complete but pregnant with wider meaning, and in the passage quoted to end this chapter there is a steady acceleration to the end of the paragraph, where there is mentioned a contrast between 'reason' and the 'geographic imagination'.

So here are the two passages one after another. My own 'explanatory' effort precedes the masterly paragraph by Forster and deliberately uses exactly the same number of words.

Erklären attempt, based on the view from spot height 199 m on the eastern Purbeck Hills (Fig. 11 is provided for those not familiar with this part of England):

The chalk ridge of the Purbeck Hills upon which we stand is continued across in the Isle of Wight, and the intervening gap is the breached and drowned southern watershed (symbolized by the remnants of The Needles) of a once-great eastward-flowing Solent River. Over the whole area that we can see, and beyond, the majority of the rivers run discordantly north–south across the east–west trending anticlines. This Hercynian trend was reactivated as the geological basement sent up

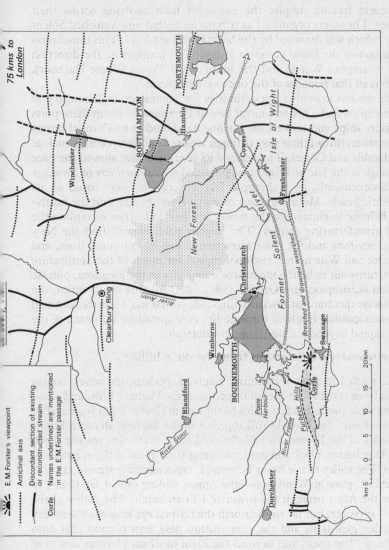

FIG. 11. *E. M. Forster's viewpoint (199 m), a climb to which is the 'wisest course' to show a foreigner England' (1910, ch. 19, opening paragraph)*

Note: The discordant sections of existing or reconstructed streams are taken from Wooldridge and Linton (1955, Fig. 19, 63). This source postulates a superimposition theory for the discordance, but later work (D. K. C. Jones, 1980, 18; Small, 1980, 68) favours antecedence. The theory that all the present streams were once tributaries of the former 'Solent River', at times when the sea level was lower than today, can be found in Everard (1954, see notably Fig. 5).

shock waves from the Tertiary Alpine orogeny. Either the present rivers are superimposed from vanished morphological surfaces or, as some now believe, they have persisted in their former direction of flow in antecedent fashion despite the east–west folds evolving across their courses. They were organized as tributaries to that now vanished Solent River, which was drowned by the latest rise in sea-level. This inundation also drowned the lower conjoint Itchen–Test (explaining the direction of Southampton Water) and the lower River Frome (Poole Harbour), which is all that remains of the once master stream.

The sea has provided opportunities for commerce and industry at Southampton and a historic naval base at Portsmouth. From these ports container ships and tankers on the one hand, and cross-channel ferries on the other, thread their way through leisure craft from centres such as the Hamble and Cowes. The sea and its pleasures have slowed the pace of change in the Isle of Wight and provided the *raison d'être* of Swanage and Bournemouth, an archetypal English seaside resort based on its crescentic beach. Motorways are creeping from the east and the north-east, helping to encourage the newer sun-belt industries to replace the older manufacturing staples. The 'higher paid salariat' find the New Forest environs and the coast attractive to live in, commute from, and retire to; and Winchester, regional capital for much of the Hampshire Basin, turns out to be the most prosperous city in the kingdom, outside London (Champion and Green, 1987, 105). All this is set within the south-west quadrant of lowland England, which has been transformed into metropolitan England through the ever-spreading influence of the great capital beyond the north-eastern horizon.

Verstehen-type description from the same hilltop:

If one wanted to show a foreigner England, perhaps the wisest course would be to take him to the final section of the Purbeck hills, and stand him on their summit, a few miles to the east of Corfe. Then system after system of our island would roll together under his feet. Beneath him is the valley of the Frome, and all the wild lands that come tossing down from Dorchester, black and gold, to mirror their gorse in the expanses of Poole. The valley of the Stour is beyond, unaccountable stream, dirty at Blandford, pure at Wimborne—the Stour, sliding out of fat fields, to marry the Avon beneath the tower of Christchurch. The valley of the Avon—invisible, but far to the north the trained eye may see Clearbury Ring that guards it, and the imagination may leap beyond that onto Salisbury Plain itself, and beyond the Plain to all the glorious downs of central England. Nor is suburbia absent. Bournemouth's ignoble coast cowers to the right, heralding the pine trees that mean, for all their beauty, red houses, and the Stock Exchange, and extend to the gates of

London itself. So tremendous is the City's trail! But the cliffs of Freshwater it shall never touch, and the island will guard the Island's purity till the end of time. Seen from the west, the Wight is beautiful beyond all laws of beauty. It is as if a fragment of England floated forward to greet the foreigner—chalk of our chalk, turf of our turf, epitome of what will follow. And behind the fragment lie Southampton, hostess to the nations, and Portsmouth, a latent fire, and all around it, with double and treble collision of tides, swirls the sea. How many villages appear in this view! How many castles! How many churches, vanquished or triumphant! How many ships, railways and roads! What incredible variety of men working beneath that lucent sky to what final end! The reason fails, like a wave on the Swanage beach; the imagination swells, spreads and deepens, until it becomes geographic and encircles England. (Forster, 1910, opening paragraph of ch. 19)

4

The Alternatives to Scientific Method in Geography Paraded in More Detail: I

Your un-neutral guide

You might reasonably ask where your self-appointed guide takes his stand. It is not difficult to deduce that my perspective contains within it a good measure of scientific method, although fuller discussion of this perspective is postponed until the last chapter. In now parading the alternatives to positivism, I shall try to demonstrate what their adherents are selling, but I cannot be relied on to be wholly fair. Advocates of particular approaches will provide what they see as the advantages of their points of view. But it is naïve to think that any one approach will provide a one hundred per cent solution to a range of problems in geography. All human endeavour involves some disadvantage, and what we seek are *net* profits of advantages over disadvantages. Because advocates of the alternative approaches are not very forthcoming about the disadvantages of what they are doing, I shall have to perform on their behalf. After all, I have quoted a formidable array of critics of the scientific method; and it is only fair to continue this wary and critical approach to all those who would like to lead geography into a more promising and 'truer' land. Series of books and papers have been produced on all these other -isms, but I have canalized my efforts here by concentrating on what geographers have written, so that forays into general philosophy and other subjects have had to be drastically curtailed.

The rise of humanistic geography

Humanistic geography surfaced strongly with a book of essays under that title edited by Ley and Samuels in 1978, but the term was first used by Tuan (1976), a paper which prompted a response by Entrikin (1976), who was among the first to point out the prominent role of humanist geography as offering a criticism of positivist geography. Entrikin actually used the term 'humanism', and, when we employ the noun rather than the adjective, we come up against a fundamental problem. 'Humanism' has many meanings, from characterizing the spirit of the Renaissance to a modern label describing those who reject all forms of religious belief. Relph (1981b) sees the danger and wishes to detach humanistic geography from humanism, proposing the term 'environmental humility':

scientism is a direct extension of the humanist principles of free enquiry and rational thought, an extension in which these principles have become dogmatic and inflexible, and therefore deny themselves. In humanistic geography humanism is challenging its own extension, not as an act of self-awareness but through the confusions that beset it and make it so difficult to know just what humanism stands for. (p. 17) . . . 'Humility' has the same etymological root as 'humanism' and preserves some of the principles of tolerance and responsibility that were to be found in the original impulses of humanism. (p. 19)

It turns out that humanistic geography is a rather wide label, and several separate -isms can be found nestling beneath it (see Pickles, 1986). It is now possible to see earlier work as being broadly humanistic, although it was not so labelled by the authors (e.g. Lowenthal, 1961; Glacken, 1967; Tuan, 1971), and the tradition has been traced right back to Vidal de la Blache and the French school of *la géographie humaine.* There has indeed been a never-submerged emphasis in human geography of giving primacy to man the actor in the dualism of man and environment, with even the environment demonstrated as having strong social and behavioural components. Pickles (1986, 1) has a preface which defends the use of 'man' as meaning humankind. Pred (1982) uses the feminine gender for his 'actress'. Pickles's defence and Pred's device will not avail me if in fact my text has sexual bias. I make no disclaimers because it appears to me as insulting to women to provide an apology on the basis that anyone could possibly believe that they could be left out of any discussion of human beings.

To be a humanistic geographer is to have considered deeply one's ontology, what Harrison and Livingstone (1980) call the 'presuppositions about the nature and reality of knowledge'. It is rather difficult to teach such presuppositions without overt brainwashing; it is far easier to learn some of the rules of the game in science which take as read certain suppositions about reality. Humanist geographers have objected to these hidden presuppositions and have reacted against the rules. We can see this by attempting to summarize some of the ideas in the Ley and Samuels (eds., 1978) volume where one cannot help encountering certain key words and phrases, and it is useful to list these alphabetically at the outset: 'context', 'contingency', 'interactionism', 'interplay of certainty and ambiguity', 'intersubjectivity', 'making man the measure', 'place rather than space', 'recursiveness', 'reflexivity'. We can now see how these arise in their context, a rather humanistic approach in itself.

The basic claim of the 1978 book is that, while the main thrust of science must be explanation (*erklären*), in human studies empathetic understanding (*verstehen*) is necessary via a self-conscious humanistic perspective. Significantly, emphasis is put on place rather than space, on man in his 'context' (a favourite word) of other men (intersubjectivity) as well as related to the natural environment, and on reflexivity: 'the capacity whereby man the phenomenon or object of understanding acts upon or reacts to the cognitive categories of man—the knowing subject' (p. 19). Another favourite word in the avoidance of physical or economic determinism is 'contingency', regarding 'man as an active agent and not simply a passive ideal type molded irrevocably by the environment in which he finds himself'. That such a self-evident claim is no longer a truism is a measure of how far man's creativity has been suppressed in geographic explanation in favour of some transcendental structure, be it 'the environment', 'the market', 'culture', or, most recently, 'capitalism' (pp. 221-2). 'Making man the measure' appears on the banner, and, with such a criterion for relevance, humanistic geography is unashamedly eclectic, catholic, artistic, and even wedded to an interplay between certainty and ambiguity. We see here a return to the holistic traditions of Vidal de la Blache rather than to the more positivist model advocated by followers of Durkheim, two schools specifically contrasted in a chapter by Berdoulay (1978, 77-90). Anne Buttimer even speculates on how Vidal de la Blache would have approached the study of the modern city (1978, 58-76).

Further light is thrown on the stance of humanistic geographers by considering what they react against. There is the complaint by

Rowles (1978*b*, 174) that in the 1960s there was so much concern with scientific precision in the use of 'uncontaminated' information that both tended to be divorced from the human situation which they aimed to illustrate. An essay by Ley (1978, 41–57) rejects the social geography of the 1960s as a meagre intellectual harvest: statistical or cartographic analysis alone proved insufficient to provide understanding of social action—back to *verstehen* again. Duncan (1978, 271) points out the danger of reification whereby man produces a world of abstractions (ideas, values, etc.) and concrete objects (the built environment, controlled nature), and then allows them to dominate him as though they were unchanging entities. He goes on to make the neat point that *dereification* occurs in the well-known process of culture shock, where all the apparatus of one's conceptual framework is called into question when playing the role of a stranger in a society different from one's own home culture.

The spatial view of geography is held by Cole Harris to fit the mobile and relatively timeless quality of American experience (1978, 123), and certainly some of his strictures about the lack of historical insight do seem particularly aimed at the excesses of the North American spatial geometers. Yi-fu Tuan advocates a bifocal vision in his analogy of the scientific method as a road towards a conclusion, while the artistic ideal is symbolized by a circle rather than by a line, by a garden rather than a path: 'If one strives for clarity and conclusion, then the ideals of completeness and closure must be sacrificed. But the reverse is not necessarily true. A road, after all, can be built within the garden' (1978*a*, 197).

Humanistic geographers have aped their positivistic colleagues in borrowing from elsewhere in the academic curriculum (e.g. physical geographers from the natural-science model, and economic geographers from economics)—a process which I should like to call 'academic translation'. Given that human geographers see subjectivity as a virtue, then there is a natural proclivity to look at what fiction writers have had to offer on the subject of place. The links between humanistic geography and literature have been actively explored (Tuan, 1978*a*; Pocock, 1981; Porteous, 1985). The words are there for the quoting, and it is taken as axiomatic that successful novelists have a heightened sensibility when describing places: remember the example at the end of the last chapter.

The translation of this sensibility into geography is more difficult when it comes to the visual arts, though it has been attempted (Rees, 1973, 1976*a,b*), and not even clouds have been overlooked (Cosgrove and Thornes, 1981). In this context I remember once rather thoughtlessly óbserving to a Dutch lady that surely she and her compatriots missed a vital visual stimulus with only those endless straight horizons for rural company. To which she poetically replied: 'Our mountains are in the sky.' And that simple statement altered for ever the way I see flat landscapes. Appleton (1975, 73 ff.) has even suggested a theory, which he calls 'prospect and refuge', attempting to explain why some views are thought to be more picturesque than others. Often the artist's experience is so powerfully transmitted as to shape a whole communal response or fashion.

Alternatively, people on the ground can speak to us. Archival work on the letters of early settlers allows us to see geography and history in the making (e.g. Birch, 1981), and, where settlement foundations post-date the written record, we may be able to demonstrate the reasons for the choice of this site over that (e.g. Kirk, 1978, site of Mandalay; Bird, 1986, site of Canberra). Thus are played out the contrasts of castle and border, townscape and wildscape, centre and periphery, or the basic human need for creating a desirable 'inscape', yet allowing a possible escape. In all this we see geographers taking on the study of the world as perceived by those who have taken the trouble to set down their impressions.

The perils of critical inviolability

A basic difficulty with humanistic geography is its fundamental emphasis on the value of subjectivity. The amount of data is enormous, and the difficulty arises in presenting results that are cumulative or hierarchical, permitting at least a few umbrella generalizations. We are warned that *verstehen* is not just empathetic understanding 'since empathetic projection could induce us to read in our own concerns to our understanding of others' (Rose, 1981, 106). So gaining access to the experience of others might be a long job. Rowles (1978*a,b*) in a study of older people's experience of a US inner city neighbourhood, took six months just to enlist his panel of participants. Then came many more months of interpersonal field-work, even though his sample numbered only five. Rowles (1978*b*, 175) points out that subjective knowing of other people is ultimately inaccessible and that objective knowing is often 'overly abstracted'. He chose a third way—interpersonal knowing: 'Emphasis is placed on interpersonal knowing through dialog rather than observation. The relationship grows beyond empathy and supersedes the exchange of individual worlds. Through "cooperative dialog" it becomes a mutually creative process' (1978*b*, 184). In Rowles's inductive procedure a hypothesis did arise in the course of his study: 'Could it be that as individuals age and actions become more constrained, there is a related expansion in the role of geographical fantasy?' (1978*b*, 185). And there was an a priori canalization of research in deciding at the outset to 'explore the geographical experience of older people' (1978*a*, subtitle), presumably to find out whether this was similar to or different from other groups.

In such studies the dangers of induction and of the too small sample are simply ignored in pursuing other benefits. Such tenets of science are widely attacked by humanistic geographers, who thereby gain some strength and cohesiveness, although they forget that much of physical geography can employ the hypothetico-deductive method without coming to much harm. Humanistic geographers seem on firmer ground when they stress the primacy of human thought in geographical studies. But because they place so much emphasis on experience and understanding, it seems hard for young geographers to achieve much until they have acquired at least half a lifetime of their own experience. Of course, the worlds of archives and artists are there for scrutiny and report but then we may commit the 'humanistic' error of wrenching the material out of its context, just like those quantifier geographers who blithely use statistics collected for one purpose to serve quite other geographical purposes.

The most extreme approach by a humanistic geographer is displayed by Olsson (1978, 1979, 1982), who has stressed the importance of ambiguity in human thought, with the word used in the following sense: 'Arising from language admitting of more than one interpretation . . . duplexity of meaning' (Webster's Dictionary). His solution has been to write geographical papers in a Joycean mode, stretching language, and often the typeface, in more than one direction at a time. His surrealist attempts, interesting at first encounter, teeter close to preciousness, and eventually promote the feeling that the experiment has been tried at least twice too often. We might argue that here we have a frustrated novelist rather than a frontiersman of geography.

But observe how one humanistic geographer deals with criticisms such as have been made above:

if ambivalence and ever-potential inconsistencies are inherent in scientific endeavour, . . . then how much more should we expect these qualities when studying man's general condition. Especially prone is work which leans heavily on the researcher's intuition and imagination and which gives centrality to subjects' intention and meaning. It is inappropriate however for these qualities to become condemnatory charges through the preemptive logic and criticism of positivist and structuralist schools, for, ultimately, different belief systems are involved. Thus, by a paradoxical twist, what critics deem foolishness, proponents proclaim wisdom: the unique may become universal, the pre-scientific may lose its hyphen to become prescience, the ragbag of approaches may be termed eclecticism and that which is of no practical value may become the most relevant. These qualities [are], then, at one both its strength yet focus of criticism . . . (Pocock, 1983, 358).

You may be able to swallow this 'paradoxical twist', but the argument is dangerous. If one argues that 'my weaknesses are my strengths', that is an attempt to silence criticism, including self-criticism. Any point of view that cannot be argued with has no means of judging its own development. We are a long way here from any hope of determining net advantages over disadvantages. On the other hand, the term 'ragbag of approaches' does seem rather harshly applied to three types of humanistic geography which do have relationships with well-known philosophies: idealism, existentialism, and phenomenology.

Idealism and geography

'Idealism' is yet another umbrella term covering a number of views which have in common the belief that reality, the world about us, is a creation of mind. As it stands this is difficult to refute, and there is always, waiting to be sprung, the idealist's trap: once any counter-epistemology is proposed, the trick is merely to say: 'Ah, that's what you think!' In practice, most proponents of the philosophical variants of idealism have not bothered to argue that material objects have no independent existence, but more mildly that such objects are not independent of minds. We shall encounter this debate again when dealing with perception, but the metaphysical discussion can be foreshortened here by looking at what geographers have made of idealistic tendencies. A point to stress at the outset is that such tendencies are anathema to those who want to believe in the power of structures and to those of materialist bent, especially historical materialists. Since these two classes of geographer have been very active in the past decade, the small band of idealist geographers have had a rough time from commentators of such persuasions.

Leonard Guelke is the name most closely associated with the advocacy of idealism in geography, and, like many geographers who can be garnered under the adjective 'humanistic', he began from a position of dissatisfaction with positivism. He even found subjectivity deep in the 'new [positivisitic] geography' of the 1960s.

The problem of testability is not solved by Haggett and Chorley [1967, 24] when they suggest that 'The terms "true" or "false" cannot usefully be applied in the evaluation of models, however, and must be replaced by ones like "appropriate", "stimulating" or "significant".' Such terms are subjective and cannot form part of deductive-nomological explanation. (1971, 47)

The new geographers ... insisted on a new methodological rigour, which attempted to eliminate the subjective element by the construction of logical and internally consistent theories and models. Yet, none of their theoretical constructs were ever complex enough to describe the real world accurately. They had achieved internal consistency by losing their grip on reality. In assessing their theoretical constructs they re-admitted subjectivity by the back door. (ibid. 51)

Guelke had discussed these dissatisfactions with Cole Harris, who had also found defects in positivism, in particular as far as historical geography was concerned; and I believe he was the first to raise idealism as a possible alternative for geographers: 'An idealist interpretation of explanation in history, such as that developed by R. G. Collingwood in *The idea of history* [1946], might suggest that behind human action lies thought, and that a historian comes to understand an event as he rethinks the thought lying behind it' (1971, 165; see also Harris, 1978, 127). Here was the seed that blossomed into Guelke's advocacy of idealism in geography (1974,

1975, 1976, 1979, 1981, 1982), holding as an 'epistemological idealist that the world can only be known indirectly through ideas' (1981, 133); that the historical geographer should have no a-priori theory of his own; and that his job was to reconstruct the theoretical framework that guided the historical actors. This of course gets more difficult in proportion to the size of the difference between the cultural context of the investigator and the action or event being studied (Hufferd, 1980, 4). Such an approach, first suggested for historical geography, was later advocated for all human geography. These are merely the headlines of Guelke's argument, but we can consider some of the ramifications in the light of the criticisms that this 'idealism in geography' has attracted.

Collingwood's *The idea of history* was edited after his death from an unfinished work on the principles of history. His view of the philosophy of history was much more complex than rethinking the thought behind past actions. For example, Collingwood believed in a dialectic between the present and the past (Daniels, 1984, 226), and the principle that 'the historical past cannot be an object of knowledge unless it is "incapsulated" within the context of the present experience of the historian. This means that in knowing the past I am also knowing the present' (Rubinoff, 1970, 298). This, and much more, is preceded in Collingwood's framework by the presuppositions of idealism in which the distinction between subject and object is overcome. Guelke merely borrowed a small part of this complex apparatus, and because he begins by being 'atheoretical' he seems to have abandoned the fundamental presuppositions of idealism in order to recommend an operation which has in fact 'A heavy reliance upon the methodology of empiricist verification' (Harrison and Livingstone, 1979, 77). By putting aside theory at the start, Guelke appears to forgo knowledge of the consequences of past actions—Collingwood's dialectic between present and past: 'It is the knowledge of the consequences of an act which alone determines whether or not it is worthy of scholarly attention' (Watts and Watts, 1978, 125). If the methodology is to rethink the thought behind the action to see the rationality of the actors' spatial decisions, then Gregory (1981*b*, 158) objects because this 'glosses over *who* is to regard them as rational and on *what* basis'. Finally, we may note how Cosgrove (1983, 100) objects to (Guelke's failure) to distinguish between conscious thought (of individuals), ideas (potentially collective via, among other media, the written word), rational understanding (logical thinking, non-conterminous with all conscious thought), and, lastly, theory (which in this context I might define as imaginative thought capable of empirical testing).

Existentialism and geography

If the main contribution of idealism to geography appears to be as a supplier of a methodology, then existentialism contributes an ontology. For the 'essence of existence' has been the concern of a group of philosophers from Kierkegaard (1813–55) to Sartre (1905–80), occupying a position opposed to the all-embracing idealism of Hegel, but finding a complementarity with materialism (especially Sartre). Existentialists stress man as the knower of his own existence and death. He is set apart from animals and a world of material objects, and strives to transcend his own existence and communicate with other existential beings. A few geographers have borrowed this perspective for a 'spatial ontology' as a struggle with a basic alienation (estrangement or failure to communicate) of man from his spatial setting. The anthropocentrism of the approach allies existential geographers to the broad aims of humanistic geography. Having set space apart—'the primal setting at a distance' (a term from Martin Buber, see Samuels, 1978, 26)—in order to confirm his own existence man must then enter into relations with space, the 'oppositeness' of which has helped to define his own essence. Animals lead subjective lives, totally immersed in their environment.

Only man requires confirmation, for only he begins in the world by defining the environment as opposite to and separate from himself. The need to confirm arises because without such confirmation, there could be no assurance that there was a someone detached, let alone any assurance that he who detaches himself has any relationship whatever to the world. Effectively, confirmation comes precisely to the extent that man exercises his will to relate or, in other words, *endeavors to mitigate distance through relationships with his environment.* (Samuels, 1978, 28)

The tension between alienation and spatial relationships is at the heart of an existential geography in the search for a higher quality of life. So existentialist geographers join in the anti-positivistic chorus to denounce the dehumanizing character-istics of Economic Man as an object for enumeration, simulation, or manipulation by mathematical logic (Samuels, 1981). Exist-entialism does lead geographers to think more clearly about loca-tions that are difficult to relate to, under the headings of 'inauthentic places', 'placelessness' (see Relph, 1976), and 'problem environments'—the problem being the breakdown of man–environment relations. Stressed also are the growing con-trasts between centres and peripheries.

The individual-centred nature of the viewpoint is obvious in statements like the following, quoted by Samuels (1978, 31) in the course of his advocacy of existential geography:

The world now without an objective center centers everywhere; and I am once more in the middle of it, though no longer objective in the sense that applies identically to everyone. The only center is the one I occupy as an existing individual. My situation is what I start from and what I return to, because nothing else is real and present, but the situation itself becomes clear to me only when I think with reference to the objective being of the world . . . I can neither grasp my situation without proceed-ing to conceive the world nor grasp the world without a constant return to my situation, the only testing ground for the reality of my thoughts. (Jaspers, 1969, 106)

Here, I think, the disadvantages of existentialism for geography begin to appear.

Most human beings have a 'home' which they regard as a 'centre' for varying proportions of the time. Yet an individual knows that this is not the centre for his community. So a continu-ous double-think is a requirement of living in a community and particularly in a community geared to an urban centre. Moles and Rohmer (1972) consider these two philosophical systems so important that they open their study of the psychology of space with a discussion of the polarity: me, here, now, a personal philo-sophy of centrality; and a philosophy of objective Cartesian space within which there are impartially observed discontinuities, like centres and peripheries (pp. 7–10). For these two authors this is 'a fundamental irreducible contradiction between two concepts of space' (p. 145; see also Tuan, 1973, 416; the above paragraph comes from a section entitled 'Centres and psychologism', in Bird, 1977a, 22–6).

We can understand that existentialist geographers would react against any 'objective Cartesian view of space', but perhaps it is being forgotten that 'space at a distance' contains the results of some of the other reactions to space past and current. Samuels seems at first to acknowledge this in the passage below, but notice how the whole system turns back to and turns on subjectivity:

Our partial spaces, in short, are rooted in and dependent upon historical and sociological situations. Though the world is always *my* place, my place has its foundations in a world apart, separate or distant from me. What I bring to the world and this place is my concern, and my concern is always contingent upon my alienation from the world. Existential space, as such, is always made in light of a correlation between my alienation from the world (objectivity) and my concern for that alienation (subjectivity). An existential geography is nothing more or less than a reflection of that correlation. (1978, 32–3)

This emphasis on subjectivity may not worry some. But there is this disadvantage. A world-view from one perspective loses tension—between my view of the world and other worlds. These are the physical material world with its own laws of cause and effect and the view of this world held by others, past and present. The richness of comparisons possible between these different worlds seems forgone by the existential approach to geography.

Phenomenology and geography

As soon as phenomenology is introduced, genuflection must be made towards the originator of this method of inquiry, Edmund Husserl (1859–1938), who as long ago as 1900 put forward an alternative to scientific rationalism. An even earlier contribution from Giambattista Vico (1668–1774) will ease us into this section: 'since men are directly acquainted with human motives, purposes, hopes, and fears, which are their own, they can know human affairs as they cannot know Nature' (Mills, 1982, 3, quoting Berlin, 1973, ii. 102). There are several difficulties involved in discussing phenomenology and geography, not least in defining the phenomenological method. Here is Relph (1981*a*), a leading geographical advocate, doing his best:

Phenomenology discloses and elucidates what we experience and how we experience it. This involves probing behind what we take for granted in everyday living and thinking . . . Our experience of things, whether illusory or substantial, transitory or enduring, are taken as the given facts which phenomenology explores . . . . It is the investigation and description of the world as we experience it directly and immediately . . . (pp. 104–5)

Many would say that a wish to provide more precise guidelines than this would be contrary to the phenomenological spirit and betray insufficient emancipation from scientific positivism. Another difficulty derives from the fact that geographers, having visited the philosophical supermarket and taken phenomenology off the shelf, have then used the product in different ways. Fortunately, the first to do so, Relph (1970), has provided an extremely succinct summary of a decade of work by geographers in this field (Relph, 1981a; see Table 11).

Relph makes the point that phenomenology is not only a method to apply to existing geographical problems, but it uncovers new perspectives that are not encountered in other mental frameworks, such as the following, expressed dichotomously: ethnocentric and egocentric spaces; back region/front region; home and journey (Porteous, 1976, 1985); authentic/inauthentic places (Tuan, 1971); topophilia (Tuan, 1974a)/placelessness (Relph, 1976); insiders/outsiders (Buttimer and Seamon, eds., 1980; Buttimer, 1983, 8). Relph (1981a, 112–13) anticipates the criticism that such topics will be thought peripheral to planning policy implications, with which geographers may be concerned, by retorting that this shows how far scientism has penetrated attitudes to the world about us, with the implication that nobody knew how to design townscapes or landscapes before the codification of 'rational' planning techniques. Perhaps we must realize that we need landscapes for different purposes at different times, for efficiency to accumulate capital to pay for those places that stimulate the mind and spirit, for man on four wheels as well as on two feet: the motorway interchange leads to the pedestrian precinct. The ideal landscape may consist of some form of labyrinth that is pleasurably solved by the user (Bird, 1977a, 146–7).

Relph confronts the problem of the absence of a datum against which phenomenological work can be judged by warning that there is no external framework (1981a, 112). The resources for such work must come from within the experience and needs of the investigator and the user. If this sounds too nebulous, let the example of horticulture stand forth, as we examine why parks and gardens are so important in advanced cultures. First, their effect is dominantly green and dendritic in pleasurable contrast to

TABLE 11. Relph's 'sequence of arguments' in the genealogy of methodological publications relating to phenomenology and geography

Sequence of arguments[a]	Relevant publications[b]
'Phenomenology is a good alternative to scientific methods;	Relph (1970); Tuan (1971)
at least, it looks useful *for some types* of geography (my italic);	Mercer and Powell (1972, citing Schutz, 1962, for human geography); Walmsley (1974, for historical and behavioural geography); Buttimer (1976) and Ley (1977) (for social geography);[c] Billinge (1977, for historical geography)[d]
well, it could be a valuable form of criticism;	Entrikin (1976)
perhaps it can be used to complement the work of scientific geographers from a humanist perspective;	Buttimer (1976); Ley (1977)
ah yes! phenomenology provides some significant qualifications within a committed scientific approach to geography'	
'. . . a concept of geography as a scientific discipline, yet one which recognises the ideological bases of knowledge and therefore requires its practitioners to be self-critical and aware of the grounds for the explanations they make'	D. Gregory (1978)

[a] Relph (1981a, 108), except for last sentence, which is p. 107.

[b] Ibid. 107–8.

[c] By 1981, Jackson was advocating for social geography a marriage of 'a Durkheimian awareness of the importance of the social with a phenomenological perspective on the contextual specificity of individual meanings' (1981, 303).

[d] Billinge is not convinced, witness this extract from his last paragraph: 'If the lessons we are to learn from phenomenology are simply that there are non-quantifiable sources which deserve and demand our attention, and that such a subjective viewpoint is not necessarily illegitimate, then these are self-evident truths which need not be hidden behind a façade of Husserlian vocabulary' (pp. 66–7).

buildings, and particularly in contrast to the modernist International Style of Architecture dominant in this century, where form reflects function with very little ornamentation. Gardens are for perambulation or for activities in a room that is demonstrably open. Formal terraces and parterres, and the herbaceous border have been succeeded by the curved lines of island beds with a constantly changing perspective, all enhanced by the march of the seasons and the fluctuating diurnal contrasts of light and weather. The garden is a theatre of growth and decay, life and death, while some of our buildings, built for longer than our lifetime, can overwhelm us by their rigid gigantism.

In contrast, wildness in nature is often a valued alternative to a 'cultivated' landscape, an antithesis cleverly exploited in Emily Brontë's *Wuthering Heights*, partially to suggest the different unconscious feelings of the main characters, one of whom is the symbolically named Heathcliff (Daiches and Flower, 1979, 145–57). The moors have been a powerful magnet for the wilderness lobby in Britain, as clearly shown by Shoard (1982). She points out that moors are found in all ten national parks, and for seven of them moorland was the reason for designation. The South Downs were not added because they were 'not wild enough'. In the US, national parks are all wild places (Shoard, 1982, 65–7). The growth of the ecological movement and 'Green' parties transforms this love of wild nature into a political force. There are changed attitudes to entire landscapes. Over the last quarter of a century Australians have ceased to regard the great hinterland of the seaboard cities as 'The Bush'. The unique flora and Australian light have been captured in many a movie; and the gardens of the newer suburbs of Canberra are planted with 'natives' while the older, inner areas sport their European 'exotics'. At the Australian macroscale, the attraction of the contrast between urbanville and a great widescape reaches a culmination in the vast, arid inland: 'For me, in a crowded world the inland contains some of the last remaining empty lands far from the madding crowd . . .' (Heathcote, 1987, 21–2).

Now for the disadvantages of phenomenology in geography. First, there are those that derive from its idealist connections, which will always call forth strong objections from geographers working in some form of structuralist or materialist framework (see next chapter). So phenomenology appears in geography like a centrist political party, ever-squeezed no matter whether the mainstream is on the left or the right. Second, the view of the individual-in-the-world underplays the way the external world may impinge on the individual. Smith (1979, 367) puts it this way: '[phenomenology] fails to take seriously the society external to the individual . . . has been loath to explain the scientific experience of reality . . . [and has an] inability to convey coherently the brutal objectivity of much everyday experience'. Third, many phenomenologist geographers have been so adept at finding new topics that they have moved far away from the Husserlian foundations. Having been 'philosophy led' to an epistemology that opened up new perspectives, they have used these perspectives to confront new problems: experiences of movement, rest, and encounter (Seamon, 1979), trajectories of individual biographies through space (Buttimer and Seamon, 1980), autobiography (Buttimer, 1983), and creativity (Buttimer, 1979). The result is that phenomenology in geography under the impulse of the recommended ontology has been dissipated into Hydra-headed, problem-led tendencies, which have blunted the total impact on the discipline as a whole.

Quite the most penetrating critique of the relationships between phenomenology, science, and geography is provided by Pickles (1985). He begins by claiming that geography remains 'ensconced in an ontology of physical nature' (p. 18):

. . . geographical inquiry is founded on an unexamined ontology of physical nature and positivistic objectivism. The resultant objectivism and epistemological subjectivism have distorted the discipline's own conception of its subject-matter and its basic concepts. In particular, they have resulted in the unquestioned adoption of a concept of spatiality most appropriate for the physical sciences, but one which is of little value in describing the spatiality characteristic of man. (p. 4).

Pickles shows how the adoption of the basic model suitable for the physical world led inexorably to the excesses of social physics (pp. 26–32). But he is also a stern critic of 'phenomenological geographers', claiming that they have either distorted or moved away from the phenomenology as conceived and developed by Husserl and Heidegger (pp. 47–67). But I find it difficult to see what a purer Husserlian phenomenology has to offer geography. It is certainly critical of 'the sedimented meanings of the everyday taken-for-granted world' (p. 170). As Entrikin (1976) pointed out, phenomenology obviously offers a critique of the scientific method, and this is still occurring far on into Pickles's text, encumbering an attempt to tell us what phenomenology 'really' is (p. 141); and even when we get to phenomenology we have not left criticism behind.

. . . the questions of whether there is a world at all, and whether it can be proved at all, and whether it can be proved and known, make no sense if they are raised by a being who is not worldless, but who is in and towards the world as its basic mode of being, and for whom the theoretical attitude [in contemporary human science] is not the most primordial mode of being.

The aim of phenomenology is therefore to clarify the mode of being of these original experiences, to make explicit the frameworks of meaning from which the sciences construct their particular thematizations of the world, and to examine critically the limits of their application as well as their relevance to the phenomena to be considered. (p. 142)

The difficulty resides in how we do this. There are only twenty-eight more pages of text in Pickles (1985) after this passage, and I have not been able to extract advice on how we should embark on the research programme he advocates in his very last words. In fact, the only practical advice I have found as to how to conduct experimental phenomenology is in Ihde (1977, 32 ff., worked out in the field of optical illusions), but, while the ontology is not difficult to understand, I cannot see how a methodology for geography can be constructed from Ihde's sequence of moves. The pure phenomenologist would say that this is a classic example of closing one's mind to what is available in the world of lived experiences.

The double hermeneutic (see also App. I.I, p. 247)

'Hermeneutics' is a word with a long history (see Gregory, 1981*a*; and Buttimer, 1983, 17 n. 6), and we can use Buttimer's definition of its modern meaning as methodological principles of interpretation and explanation. We have seen how humanistic geography, and phenomenology in particular, enjoins the student to bring to a study of any aspect of society a sympathy for the values it enshrines and an understanding of how it views the world and then interpret that interpretation against some datum, which will include the student's own hermeneutic—hence the double hermeneutic of Giddens (1976, 162), because human geography, like sociology, 'deals with a universe which is already constituted within frames of meaning by social actors themselves, and reinterprets these within its own theoretical schemes . . .'. Schutz (1967) has attracted attention in this respect because of his attempt to avoid the transcendental element in Husserlian phenomenology (see Gregory, 1978, 134 ff.); and Ley (1977, 498) even believes that Schutz's theory of social action 'is . . . an appropriate underpinning for social geography'. Schutz's 'dual vision' has been summarized by Gorman (1977, 138–9):

The scientific method appropriate to studying society describes second-order ideal type constructs, modelled after the primary idealizations actually influencing social actors, and explaining typical, regularly performed interaction. These ideal type models, the homunculi, when empirically verified, explain this interaction scientifically while simultaneously remaining true to actual meanings experienced by actors in typically encountered situations.

But Gorman goes on to point out that if you ground objective knowledge in empirical reality, as Schutz does, you cannot at the same time uphold consciousness as the ultimate source of all knowledge, which Schutz also does (Gorman, 1977, 140). Another grand compromise collapses.

An interesting illustration that the double hermeneutic throws on the concept of 'net advantage' of one philosophy or methodology over another is provided by Giddens (1976, 144; also reported by Gregory, 1978, 144): 'The process of learning a paradigm or language-game as the expression of a form of life is also a process of learning what the paradigm is not: that is to say, learning to mediate it with other rejected alternatives, by contrast to which the claims of the paradigm are clarified.' It is a great pity that the word 'paradigm' surfaces again and also that the 'alternatives' have to be unceremoniously dubbed as 'rejected': there are surely stages of selection before complete rejection, just as there are horses for courses. But how telling is Giddens's example of the frame of reference called Protestantism, which cannot be fully understood without a study of the frame of reference against which it originally protested (see Fig. 12).

This chapter appears to have led us far from physical geography, but perhaps the double hermeneutic, operational in human

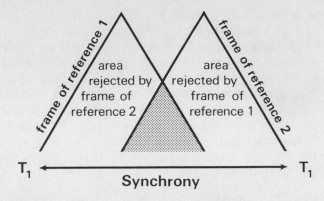

FIG. 12.   *Two frames of reference at any one time*

Two examples: one society compared with another via a different mix of components (Dumont, 1970, 231–4; Fig. 5; and 340 n. 118d); one national urban system superimposed on another for comparative purposes in a 'merged network' (L. S. Bourne, 1974, 166, Fig. 6, bottom). Such comparative approaches when applied to methodology might not preclude preference for one particular form of inquiry in a state of temporary net superiority for a particular purpose or scale (the idea of 'horses for courses'). A list of methodological mechanisms involving comparison is given in Table 30.

Source: Bird, 1979, 121; cf. Fig. 1 in this book.

geography, can at least point up a contrast. Surely the only hermeneutic in question within physical geography is that of the investigator. The phenomena being studied have no minds of their own in a universe where rule the physical laws of cause and effect. Caution! If the result of man's actions form a significant part of the physical system being studied, or if it is wished to apply the principles of physical geography to a particular area, then the double hermeneutic comes racing in. The man in the 'man's actions' and the users of a particular area may have frames of reference completely different from that of the scientific investigator. These frames of reference may be scientifically wrong, but even a wrong hermeneutic is nevertheless relevant to a full understanding of that area's physical geography (see Clarkson's 'four realities', 1970, Fig. 3, 715).

Feminism and geography

This section appearing as a coda to the chapter may seem like tokenism in the direction of a sensitive area. In fact, this position is symbolic because the topic 'geography and gender' illustrates well the contrast between humanistic geography and more radical approaches. Geography has a flank open to the currents of all social movements; but even if it were merely a science of space, one could easily demonstrate that the territories and orbits of women differ markedly from those of men in most societies (Everitt, 1976, Figs. 2 and 3, 109–10; Bird, 1977*a*, Fig. 23, 142). 'Patriarchy is a concept which soon arises in feminist studies: a set of social relations between men which, although hierarchical, establishes an interdependence and solidarity between them which allows them to dominate women' (*Geography and Gender*, 1984, 26).

As a result of 'the first explicit discussion of a central feminist concept in a major geographical journal' (Johnson, 1987, 210, referring to Foord and Gregson, 1986), there is a debate as to whether gender relations and mode of production are separate objects of analysis (Gregson and Foord, 1987, 373), or whether patriarchy is to be explained by a Marxist analysis of class relations (McDowell, 1986, 312; Gier and Walton, 1987; Johnson, 1987, 213; see also Knopp and Lauria, 1987). But all feminist geographers seem to agree on the central importance of male domination over women. Surely investigations by humanistic geographers would have revealed the spatial effects of this.

. . . the contributions from phenomenologists and humanists on woman's place have been meagre . . . In fact, it could be argued that . . . [they] are basically modest reformers, looking for ways in which the ordinary person may better their lot within the chinks and interstices left between the great power relationships in society, adjusting to the prevailing patterns of domination and subordination. We should add here that phenomenologists and humanists tend to show a *general* concern for the ways in which ordinary people are subject to various forms of authority, rather than analysing the specific forms of exploitation and oppression that occur. Their insights are, therefore, of limited value to those interested both in analysing the specific forms of oppression to which women are subjected, and in using this as a basis for removing such oppression. (*Geography and gender*, 1984, 36–8)

The earth has moved in this passage. Explanation (*erklären*) and empathetic understanding (*verstehen*) are not enough. The aim now is to change society via 'praxis', which has been defined by Wisdom (1987, 81) as 'a post-Marxist doctrine meaning roughly that all genuine theoretical work must result in social application'. And so feminism is divided into 'radical feminism', 'socialist feminism', and 'marxist approaches' (*Geography and gender*, 1984, 26–38); right-wing feminism is not very visible. This political dimension will surface even more prominently in the next chapter.

What feminist studies teach is the power of gender-dominated mental frameworks; the role stereotyping of men and women is very powerfully imprinted within us all. One curious feature of this whole subject is the too sudden, often angry or dismissive response to feminist geography from many male professional geographers, the strength of the reaction serving to confirm the rigid mental compartmentalism to be overcome; yet long ago Jung taught us about the archetypal anima in men and its male counterpart, the animus, in the female psyche. An example of a statement that can be taken as one of the 'givens' in our view of women's role in society now follows; and this comes from the first feminist book on geography and gender! 'Women's ability to take on waged employment is dependent on managing their domestic duties as well' (*Geography and gender*, 1984, 84). Yet from this simple statement two geographical topics of great importance derive: the importance of female labour in society, its distinctiveness, and spatial segregation; and the dependence of women on access to certain social services which enable them to be employed (child care, health services, public transport, legislation *re* maternity leave). If these enabling conditions are differentially provided over space, then the whole fabric of female opportunities for employment is distorted. Transport in the city is one area where women and geography come together, because on access to transport depends access to services and employment, and even personal safety. Women are much less likely to have access to a car, twenty-four hours a day—they may not even be allowed to drive the one family car. Use of public transport exposes them to greater risks of violence, particularly when walking or waiting in hours of darkness (see *Women on the move*, 1987). Yet few texts on social geography make special reference to spatial problems faced by women, and it is fair to say that women have been hidden from geography, just as they have been hidden from history (Foord and Gregson, 1986,

188 and 207 n. 4). Finally, the reference to domestic duties gives a distinctive bias to environmental perception by women: 'home' is a symbolic datum of subtly different weight for women compared with men (Mazey and Lee, 1983, 37–47).

Feminist geographers offer new topics and problems for study, but the approach, perhaps through proselytizing zeal, has the disadvantage of being capable of including extreme statements that alienate the audience that is the target for conversion. Three examples:

While orthodox transport policies continue to advocate environmentally damaging car ownership and use, many women will continue to experience driving as liberating, freeing them from an ill-adapted and unsafe public transport and urban environment. Until such time as transport policy reflects their needs, women must have the choice between driving and other modes in the same way as many men do. (*Women on the move*, pt. 10, 17–18)

The 'until such time' suggests that, once public transport is adapted to their needs, women with access to cars would give up driving.

To some extent, patriarchy works because men at upper levels in the hierarchy can 'buy off' those at lower levels by offering them control over at least some women. (Hartmann, 1981, quoted by *Geography and gender*, 1984, 26)

Put at its crudest, the men go out to work and are exploited there by the capitalist, and then return to their homes and families where they in turn subordinate and oppress 'their' women. (*Geography and gender*, 1984, 30–1)

Feminists stress gender differences rather than biological differences, and this tends to play down or even ignore a basic biological role in reproducing society and the values it enshrines, including incidentally a share in the responsibility for gender stereotyping. Even more fundamental is the role of women in giving long-term continuity to a society. Warring factions of men can dominate the headlines, but it is often the women who ensure that life goes on in even the most troubled societies. This was recognized in *The grapes of wrath*, a novel which geographers often turn to for its graphic description of the literally *man*-made 1930s catastrophe of the American Dust Bowl. But the basic theme of the novel was the strength of the life-force (centre-piece of the philosophy of Henri Bergson, 1859–1941, popularized by G.B. Shaw). The novel shows how a group of migrant workers copes with appalling hardships, yet begins to put roots down in the new area. The following speech of Ma Joad, beautifully delivered by Jane Darwell in the John Ford film version, illustrates the continuity that women can give to a society:

Women can change better'n a man . . . Woman got all her life in her arms. Man got it all in his head. Don' you mind. Maybe—well, maybe nex' year we can get a place . . . [Pa interjects that it seems that their life is over and done.] No it ain't . . . It ain't, Pa. An' that's one more thing a woman knows. I noticed that. Man, he lives in jerks—baby born an' a man dies, an' that's a jerk—gets a farm an' loses his farm, an' that's a jerk. Woman, it's all one flow, like a stream, little eddies, little waterfalls, but the river it goes right on. Woman looks at it like that . . . (Steinbeck, 1939, 388)

It may be stretching things a bit to say that lurching from one -ism to another is a rather active 'masculine' feature, and that the deeper, longer-term changes in geography represent its truer lifeforce, a dynamic curiosity involving the world about us and a 'feminine' sympathy with the rest of the biosphere. Such 'stretching' analogies are found in humanistic geography, for good or ill, or a combination of both.

5

The Alternatives to Scientific Method in Geography Paraded in More Detail: II

Structure as stricture

> **Stricture** ... The action of binding or encompassing tightly.
>
> *Oxford English Dictionary,* extract from meaning 2)

One can readily understand why geographers have become interested in structures, and this includes the turn towards systems in physical geography. The desire to answer 'how?' questions, the interest in processes, develops into 'why?' questions in human geography, and ultimately into 'what for?' and 'for whom?' debates. Research benefits from comparison with a datum, and this can be provided by a definition such as 'optimum allocation of resources' provides for economics. This does not tell us how the optimum is to be determined, nor who does the allocating. So we still need to know the structure in which the process takes place. The structuralist, if for the moment you will permit this very generalized being to step forth, says that behind the world of appearances lies a world of structures, and these, hidden at first, can be revealed by logical thought, and tested against how the system works. A danger looms. If a structure is posited, it could remain in place even as the world changes, though another structure, over-arching at a higher hierarchical level, could allow one structure to be transmuted into another. Yet the approach relies on closure: sooner or later the threatened millennium will arrive. As we have seen with the hypothetico-deductive method, there are tremendous advantages in theory-laden observation, in having a framework of thought at the outset, provided that this is abandoned if a better theory supervenes—better in that more of the data are explained. It does not seem to be in the nature of approaches that can be broadly labelled 'structuralist' to contemplate erecting a structure which contains the possibility of its own collapse, unless of course the 'over-arching' system allows another structure to arise as an immediate phoenix; and if that collapses, yet another structure . . . After all, with no structure, no structuralism. We will quickly understand the power conferred when thinking structurally, but in that very power lies danger. Some have retreated from thoroughgoing structuralism into another order of processes: where the structures are recognized by those within them, who made them in the first place, and who now take them into account when acting out their individual roles. But, even in these more recent versions of structuralism, the structures are still hovering over the scene, like the Eumenides, waiting to pounce on the actors.

The legacy of Karl Marx

Two giants of the nineteenth century seem with us still: Darwin and Marx. Both promulgated metatheories, the status of which look relatively different over a century later. Darwin's theory needed a mechanism, later provided by Mendelian genetics, and encountered a rival, Lamarckism (and a later version called Lysenkoism), which it has defeated in every test so far. Darwin had the advantage over Marx in working in the non-social field of biology. Any theory in the field of social science carries the seed of its own supersedure, because the social actors to which the theory refers can act in the light of the theory once published. But there are other difficulties in considering the work of a nineteenth-century social scientist, even if he was one of the world's greatest thinkers (see Table 12).

As a result of this contrast, it should come as no surprise that, whereas Darwinism has merely been succeeded by neo-Darwinism, Marx's original formulation has spawned a host of successors:

1. Scientific Marxism (Gouldner, 1980, 32–63; and for adherents see p. 38);
2. Marxian economics ('antipathetic to capitalism; whereas the neoclassical [economic] perspective provides ideological support for the competitive free enterprise system', D.M. Smith, 1981a, 203);
3. Marxist humanism (Baker, 1981; Gregory, 1981c);
4. Marxism-Leninism (via Stalin);

TABLE 12.   Possible errors (now apparent) made by a social scientist working during the nineteenth century

May now be judged wrong because:

1. Studies societal change up to nineteenth century, but society has has changed further since then[a]
2. Changes may have occurred because of existence of social science studies, including his own[b]
3. May have misinterpreted data through faulty logic[c]
4. May have used what are now thought to be wrong or inadequate data[d]
5. May have been ignorant, in common with others, of correct data[e]
6. May have been ignorant of techniques now available[f]
7. May have failed to present net advantages of alternative hypotheses in competition with chosen theory[g]

Examples of above errors in the case of Marx:

[a] Class was a dominant feature in nineteenth-century Britain. But, even confining the 'universe' to Britain, class, gender, language, nation, religion have all undergone significant absolute and relative changes since the time of Marx.

[b] Think of all the 'Marxist studies' since Marx, and actions in the name of Marxism, and its variants.

(i) Generally—game of Cheat the Prophet. If a prophecy is made, after the death of the prophet a group could conspire to act in defiance of the prophecy (the opening of the novel *The Napoleon of Notting Hill* by G. K. Chesterton, 1904; called 'the Oedipus effect' by Popper, 1961, 13 and 15–16).

(ii) In particular—if a prophecy includes bad news for an individual or a group (capitalists in the case of Marx), forewarned, they can take steps to thwart the logic that leads to the prophecy.

[c] Human fallibility, and 'Marx updates British economic statistics when its suits him but retains the older figures when they support his case' (Elster, 1986, 188).

[d] Applies to Marx's work on the Asiatic mode of production (ibid. 187).

[e] 'What accounted for the marked rise in urban population in the eighteenth century was not expropriations from the land as a result of enclosure [as in Marx's argument] but the fact that in the eighteenth century, there was an enormous increase in population in town and country alike' (Conway, 1987, 123).

[f] . . . analytic tools for the study of monopoly did not exist at his [Marx's] time' (Elster, 1986, 187).

[g] 'Certainly there is no trace in his writings of the scholarly practice known as playing the devil's advocate' (ibid. 188).

Note: P. J. Taylor (1987, 305 n. 7) also realizes that he is treating Marx as 'a man of the nineteenth century' while 'by no means disparaging his contribution to socialism'. Taylor sees the nineteenth-century struggles of socialism as centred on nation states,' whereas [in modern politics] the social and economic forces that are continually changing our world are manifestly global in scope . . .' (p. 287).

5. Critical Marxism; or humanistic Marxism; or neo-Marxism (Frankfurt School—from Horkheimer to Habermas), paying more attention to the superstructure and advocating more open-endedness, though still predicting revolutionary change (see Gouldner, 1980, 38 for adherents; Keat and Urry, 1975, 218–27; and D. Gregory, 1981*d*);
6. Polycentrism (Castroism, Maoism, Titoism).

Some of these variants have provided the philosophical basis for work in human geography; others have caused wholesale changes in national landscapes. But, in case you are not sure what all the fuss is about, here is one of the most famous formulations— Marxism by Marx in fact, in the preface to *A contribution to the critique of political economy*:

In the social production of their life, men enter into definite relations that are indispensable and independent of their will, relations of production which correspond to a definite stage of development of their material productive forces. The sum total of these relations of production constitutes the economic structure of society, the real foundation, on which rises a legal and political superstructure and to which correspond definite forms of social consciousness. The mode of production of material life conditions the social, political and intellectual life process in general. It is not the consciousness of men that determines their being, but, on the contrary, their social being that determines their consciousness. At a certain stage of their development, the material productive forces of society come into conflict with the existing relations of production, or, what is but a legal expression for the same thing—with the property relations within which they have been at work hitherto. From forms of development of the productive forces these relations turn into their fetters. Then begins an epoch of social revolution. With the change of the economic foundation the entire immense superstructure is more or less rapidly transformed . . . In broad outlines, the Asiatic, ancient, feudal, and modern bourgeois modes of production can be designated as progressive epochs in the economic formation of society. The bourgeois relations of production are the last antagonistic form of the social process of production—antagonistic not in the sense of individual antagonism, but of one arising from the social conditions of life of the individuals; at the same time the productive forces developing in the womb of bourgeois society create the material conditions for the solution of that antagonism. This social formation brings, therefore, the prehistory of human society to a close. (Marx, 1962, i. 362–4)

What we have here is essentially a materialist conception of history, often referred to as historical materialism. This is why all who subscribe to this family of approaches are strongly opposed to any form of idealism. The materialist element resides in the economic basis of all modes of production, and the structural element in the concept of the political and ideological superstructure erected on this basis. Notice also that capitalism appears as the last stage before the millennium, which helps to explain why capitalism is the dominant target of Marxist studies. This legacy of Marx is very much with us in present-day geography. Three quotations will give a flavour of how historical materialism appears in the subject, first in respect of the relations of man and environment:

Man's relationship to nature and to his fellow man are relationships that are intrinsically social and historical. Any approach that does not recognise these inherent qualities will, therefore, fragment into a barren descriptive subjectivity . . . concepts of nature are historically specific (i.e. appropriate to a given historical and social configuration) and . . . changes in such concepts correspond to changes in the way man organises his way of interacting with nature. (Burgess, 1976, 21–2)

Second, from a recent overview of work in historical geography, we can see the unchanging Marxist view of capitalism: 'Central to the Marxist theories is the distinctive mode of production labelled capitalism, with which is associated deprivation, commodification, proletarianisation and immiseration of the majority of the populace, agricultural and industrial, which come under its sway . . .' (Butlin, 1987, 29–30). And, finally, here is a succinct definition of a historical materialist geography, which in fact is put out as a definition for the entire discipline: 'Geography is the study of spatial forms and structures produced historically and specified by modes of production' (Dunford, 1981, 85).

Historical materialism in geography, including Marxist geography, welfare geography, and radical geography

A strong feature of historical materialism in geography is an anti-capitalist stance—against the free action of market processes and the emergence of a hierarchy of classes in the processes of production. Radical geography is overwhelmingly historical materialist, 'left-oriented' (Owens and House, 1984, p. ii). But 'radical' is neither a wholly accurate nor a positively coherent term: it is inaccurate because in anti-capitalist societies a radical geography would be pro-capitalist; and, second, 'Radical geography . . . has only a negative coherence—its opposition to non-radical geography' (N. Smith, 1979, 374–5). Indeed, even 'Marxist geography' is not a particularly happy term, first, because it does not specify which detailed variant of Marxism is being employed as a theoretical foundation; and, second, and much more importantly, Marxism is a holistic approach applied to every aspect of the development of society which cannot be applied piecemeal to aspects of society such as spatial patterns emanating from societal processes. There is a basic conflict here. True to its Marxism, 'marxist geography would transcend disciplinary boundaries, true to its geography it would observe them' (N. Smith, 1979, 376). Those who incline to the first part of this dichotomy as a purer Marxist approach have dubbed those who have slid into the second part as indulging in spatial fetishism (N. Smith, 1979, 375; see also Quaini, 1982, 157–8). Nevertheless, an injection of historical materialism does throw a distinctive perspective on geographical problems, and gives geography a political purpose: 'geographers inevitably practise politics whilst politicians just as inevitably engage in geographical practices. Marxists freely acknowledge that link and seek to strengthen rather than hide it' (Harvey, 1981, 210). But Harvey would go further: 'The Marxist project is a revolutionary project in the broadest possible sense. The aim is not merely to understand the world but to change it' (1981, 209).

Historical materialists believe Western-style capitalism to be currently crisis-ridden, eventually collapsing through the basic inconsistencies of its inherent structure. These crises can be studied by geographers as they work themselves out over space (Table 13), leading to the final revolutionary outcome when communism will succeed capitalism, although this outcome, originally predicted by Marx more than a century ago, seems ever-delayed.

Among the more obvious reasons for this has been the growing power of organised labour to bargain with capital for increases in real wages [= gains of socialism delay communism?], which capital has been able to concede to some extent by virtue of the great success of the capitalist system in expanding the capacity to produce. It may also be the case that rising affluence, at least in the advanced capitalist world, has helped to diffuse working-class consciousness—a trend assisted by use of the mass media to stimulate materialistic values and reinforce the prevailing ideology of capitalism. (D.M. Smith, 1981a, 208)

TABLE 13.   A radical geographer's crises of capitalism[a]

Cause	Crisis and spatial arena
Process of capital production	Class inequalities, predominantly racial in US; persistent inequality erupting into social and spatial problems in US
Imperialistic use of Third World by First World	Wars of liberation in Third World, and escalation of price of energy materials there,[b] injuring First World
Capitalist systems' need to expand production constantly; destructive commodity attitudes towards environment	Resource shortages and pollution of regional and national environments
Interaction between elements in, above, and between each cause	Heightens each crisis and produces hybrid crises leading to escalating social crises in capitalist countries[c]

[a] Peet (1977, 1). For a book-length critique of 'radical development geography' see Corbridge (1986).

[b] 'It is a central tenet of Marxism that the end of capitalism has been postponed by exploitation of the people, markets and resources of Third World countries' (Peet, 1977, 3).

[c] '. . . contradictions are nurtured by the process of capital development, are enhanced by the growth of the system, until they erupt in the form of a seemingly endless series of irresolvable crises (crime problems stemming from inequality, regional alienation from cultural oppression, etc.)' (ibid.).

An important feature of capitalism is seen by Marxists as having notable geographical effects—the unequal accumulation of capital, with capitalists having it and workers not. Moreover, these two classes are seen as spatially segregated, leading quickly to the idea of unequal spatial development at all scales: inner cities versus suburbs, unbalanced regional development, advanced capitalist world and less developed world. The existence of this last contrast spawns the idea of a neo-colonialism, of exploiters and exploited. Study of uneven development was once promoted by D. M. Smith as 'welfare geography' (1977), perhaps too glibly characterized as 'who gets what and where'. But this approach is now considered by Smith himself as too descriptive and to have given way to process-oriented work (1981b, 371). He concludes: 'Implicit in "welfare geography" is a recognition that the issues in question extend beyond the limits of a single discipline, and in fact render disciplinary boundaries increasingly irrelevant. The welfare approach logically requires a holistic social science perspective' (p. 372). Guess what that might be.

It is held by historical materialists that the dynamics of capitalism compel the system to over-accumulate capital in relatively few hands, exacerbate class differences, and increase spatial inequalities. The growing power of multinational corporations enables them to transcend national state boundaries and to use cheap labour wherever it can be found: for example, high quality, advanced, technical goods assembled in the poorest countries. Thus ever-larger capitalist monopolies confront increasingly impoverished working classes. The over-accumulation of capital in relatively few hands superheats technological development leading to crisis after crisis, with the entire, pilotless system manifesting an uncaring approach to the natural environment, ravaging its resources for ever-greater short-term gains.

The political implications of the perspective are obvious, and historical materialism has a wide programme to offer in geography, and many are engaged in promoting this with proselytizing zeal. According to Harvey (1981, 209), there are three journals that do just this: *Antipode* (US), *Hérodote* (France), and the *International Journal of Urban and Regional Research* (Britain); while the agendas of three classified bibliographies of Marxist (radical) geography (Barton, 1980, 1983; King, 1982; Owens and House, 1984) have altogether sixty-three subsections. These can be reclassified into the following eight broad areas of concern:

1. Concept of Marxism as an attack on 'mainstream' geography in Western-style capitalist countries,
2. Role of classes in the State in Western-style capitalist countries,
3. Urban analysis,
4. Minorities and social inequalities,
5. Housing,
6. Rural and environmental studies,
7. Regional inequality in Western-style capitalist countries ('Work on regional inequality in socialist countries is rather limited', King, 1982, 188),
8. Dependency, imperialism, colonialism (intranational and international).

Even this secondarily derived and highly generalized agenda

shows the emphasis on the concern to probe inequalities over space as the result of the working of Western-style capitalist economies.

The foregoing summary of all this activity is obviously very short, but even so one feature is revealed quite clearly. Just as 'mode of production' is a vital foundation of Marxism, so an anti-capitalist perspective characterizes all historical materialist work in geography:

> The be all and end all of struggle for me right now is capitalism. There are all kinds of other struggles going on which are important, but the fundamental one that I am concerned with is the struggle against capitalism, precisely because that lies at the centre of the most virulent forms of oppression at the same time as it poses the most serious threat to the continuation of any kind of civilization. (Harvey interviewed in Peake and Jackson, 1988, 17)

Non-Marxist approaches are dubbed by radical Western Marxist geographers as supportive of the status quo, and riddled with unstated ideological assumptions: 'those who stick to the abstract-materialism of natural scientific method and who locate their modelling efforts within that frame are the true ideology producers in contemporary geography' (Harvey, 1988). It would be understandable if Marxist geographers felt themselves as likely to be persecuted in what they see as the hostile system in which they work. This has certainly not happened in the case of David Harvey who has held chairs in both the US (Johns Hopkins University) and, currently, in Britain (Oxford). This leading advocate of the 'Marxianization' of geography is particularly remarkable because in 1969 he published a book which has often been held to symbolize the apogee of a scientific approach in geography.

David Harvey in the literature

Here is an intriguing case. In *Explanation in geography* (1969a) Harvey advocated the deductive-predictive approach; yet just three years later he published a paper (1972a) which showed he had embraced Marxism; and he has been a leading Marxist geographer ever since. These are just simple headlines, and, at the other degree of detail, a book-length treatment of Harvey's writings up to 1981 has appeared (Paterson, 1984). This was not written from a Marxist perspective but in the light of 'non-relativistic pluralism':

The integrity of each philosophical perspective ought to be recognised and allowed to develop freely within the discipline. This need not imply a philosophical relativism, that is, that each approach is as 'truthful' or valid as any other. Rather, a non-relativistic pluralism entails critical commitment to that perspective which the researcher finds most in accord with his or her view of reality. (p. 175)

This perspective of Paterson has violently upset Marxist geographer, Eliot Hurst (1985, 70–1), for historical materialism does not permit such permissiveness. Harvey's view seems more in favour of pluralism (1977, 407), possibly because, as Paterson suggests (1984, 108), this allows Marxist geographers to work without fear of repression.

Since we know that Harvey's first Marxist publication was in 1972, the following questions arise:

1. What was going on in Harvey's mind from 1969 to 1972? (leading to:)
2. What caused him to turn to Marx?
3. Are we justified in saying that from 1972 to date, Harvey has subscribed to the historical materialism that was described in the last section?

To help answer the first two questions we have the evidence of what Harvey published in the period from 1969 to his first Marxist paper in 1972a (a full list of these publications is in Paterson, 1984, 179–80). In 'Conceptual and measurement problems in the cognitive-behavioral approach to location theory' (1969b), he surveyed normative location theory, stochastic location theory, economic rationality, and satisficing behaviour; and this led to perception studies where a firmer theoretical framework was needed. In a book review of the same year, he urged that we should pursue our studies 'with much greater rigor and clarity' (1969c, 314). From 'Social processes, spatial form, and the redistribution of real income in an urban system' (1971) comes this in the last paragraph:

I have concentrated my attention upon the mechanisms governing the redistribution of income and I have suggested that these seem to be moving us towards a state of greater inequality and greater injustice . . . I . . . conclude that it will be disastrous for the future of the social system to plan ahead to facilitate existing trends . . . I . . . find the notion that we are moving in easy stages into an era of enormous affluence and electronic bliss unacceptable since it is at variance with my own analysis and the evidence of my own eyes . . . We really do not have the kind of understanding of the total city system to be able to make wise policy decisions, even when motivated by the highest social objectives. It seems, therefore, that the formation of adequate policies and the fore-casting of their implications is going to depend for their success upon some broad interdisciplinary attack upon the social process and spatial form aspect of the city system. (p. 298)

Notice in this passage: 'my own analysis and the evidence of my own eyes'. This refers to his experience of the Baltimore housing market after taking up an appointment at Johns Hopkins University in 1969, when he was no doubt struck by the stronger materialistic basis and inequalities of the American way of life compared with the British. In the above passage there is also the yearning for 'some broad interdisciplinary attack' on the problems he was encountering.

Ollman's (1971) book on Marx must have been an influence with its emphasis on the philosophy of internal relations that Ollman saw as the heart of Marx's method. This later surfaced in Harvey's (1973) *Social justice and the city* (pp. 288–92) and in a paper on the Baltimore housing market (Chatterjee, Harvey, and Klugman, 1974), both with specific acknowledgements to Ollman:

No one would deny that things appear and function as they do because of their spacial–temporal [*sic*] ties with other things, including man as a creature with physical and social needs. To conceive of things as Relations is simply to interiorize this interdependence—as we have seen Marx do with social factors—in the thing itself. (Ollman, 1971, 27–8)

We could allow our categories and meanings to change coherently and relationally in ways which mirror the changing existence we seek to portray . . . As Ollman has recently pointed out in a most perceptive book, this is exactly what Marx does in a calculated and rigorous way. (Harvey, 1972*b*, 327)

Of course, Harvey also began reading Marx:

At the bottom of it all, Marx argued, lies a tension between capital and labor . . . We could, Marx suggested, create a theory to explain the contradictions at the same time as it would help us overcome them. I found this very exciting and started to work at it. And lo and behold, one by one, the contradictions which had so perplexed me crumbled before the power of the analysis. (Harvey, writing in the *Baltimore Sun*, 14 May 1978, and quoted by Paterson, 1984, 145)

So we have the following strands in Harvey's philosophical metamorphosis: (1) the arrival in America in 1969; (2) the evidence of great inequalities in the Baltimore housing market; (3) the search for 'some broad interdisciplinary attack' that had as much 'rigor and clarity' as the deductive-nomological model of scientific explanation he had advocated in *Explanation in geography*—an attack which would be more relevant to the ultra-capitalist system in which he found himself; and (4) the influence of Ollman, who stressed the relational aspect of Marx's writings. Thus was prepared the ground for a receptive reading of Marx, leading to Harvey's first Marxist publication in 1972, 'Revolutionary and counter-revolutionary theory in geography and the problem of ghetto formation' (1972*a*). This paper started with Kuhn's (1962) analysis of revolutions in the natural sciences and H. G. Johnson's (1971) recipe for a successful revolution in an academic subject, and then turned to the reasons why counter-revolution might arise in a social science. This was because social science could not be divorced from the society it studied, and a revolutionary theory in social science could only gain acceptance if its reading of social relationships could be seen working in the society at large. A counter-revolutionary theory is one that is proposed to prevent the social changes which adoption of the revolutionary theory would bring about (this is expanded on p. 41 of the paper). Turning to geography, he suggested that we could abandon positivism for idealism, phenomenology, or a naïve positivist empiricism, and that the so-called behavioural revolution in geography had flirted with all these approaches. Then comes the passage in which Harvey first brings Marx into geography:

The most fruitful strategy at this juncture is therefore to explore that area of understanding in which certain aspects of positivism, material-ism, and phenomenology, overlap to provide adequate interpretations of the social reality in which we find ourselves. This overlap is most clearly explored in Marxist thought. Marx . . . gave his system of thought a powerful and appealing phenomenological basis. There are also certain things which Marxism and positivism have in common. They both have a materialist base and both resort to an analytic method. The essential difference of course is that positivism seeks to understand the world whereas Marxism seeks to change it. (pp. 6–7)

Next comes an argument on urban land-use theory, followed by identified 'manifestations of this general condition in the urban housing market' (pp. 9 ff.), which no doubt derived from his immediate experiences of studying conditions in Baltimore. In his last paragraph we find that 'the emergence of a true revolution in geographic thought is bound to be tempered by commitment to revolutionary practice' (p. 11).

Harvey has himself compared his present Marxist perspective with his pre-1972 deductive-nomological framework:

I was recently asked to think about the *Models in Geography* volume [Chorley and Haggett, 1967] which came out 20 years ago. I read the first two pages of the article [Harvey, 1967] I wrote for that and I still agree with them. Basically, what I was saying is that all geography has to be historical geography. We have to have an understanding of how space and time are being used to create historical geography in society. What I wanted, however, was a way of pursuing that that was going to be powerful and rigorous. The reason I got intrigued with the quantitative, positive movement was that I thought this was the way to go. The reason I got disillusioned with it was it didn't seem to be producing anything in the way of results. So I came out of that and the switch into Marxism was really intellectually a way of saying, well, finally I've found something that seems to work in relationship to what I was trying to do all along.

. . . I wanted some kind of intellectual understanding of what was going on that was going to be consistent both in the project of understanding historical geography and also consistent with the notion of looking for a knowledge system, if you like, which was about empowerment of disempowered people.

. . . It was a political transformation as well as an intellectual one. (Harvey, interviewed in Peake and Jackson, 1988, 11–12)

Turning to what has happened to Harvey since 1972, we can observe that he certainly began as a scientific Marxist. We see him immediately 'inverting' the Adams–Wheatley theory of urban origins in a review dated 1972. This theory finds religion, a 'ceremonial complex', as the *primum mobile* of the origins of the earliest cities. Using Wheatley's own evidence, Harvey's 'perspective is to seek a materialist interpretation along the lines suggested by Marx' (1972c, 510). This contrast of theory is summarized in Table 14. Harvey continued in this scientific Marxist mode for some time, and to that extent our third question is answered: he did subscribe to something very close to the historical materialism outlined in the previous section.

TABLE 14.   Thesis and antithesis in the founding of the first cities

Adams–Wheatley sequence[a]		Harvey sequence[b]	
Pre-1 (S)	Egalitarian society	Pre-1 (S)	Egalitarian society
1 (P)	Religious movement	1 (P)	Material mode of existence transformed economically
1a(P)	Secular forcès assist		
2 (S)	Redistributive and rank society based on effective space (umland) giving surplus	2 (P)	Transformation in consciousness due to 1
2a(P)	Surplus maintained by stable peasantry attached to religious image projected from a centre	2a(P)	Change in ideology
		3 (S)	Redistributive economy and rank society
		3a(P)	Surplus produced by a peasantry with low migration potential

S = Stage, P = Process

[a] Adams (1960a, 1960b, 1966); Wheatley (1971).

[b] Harvey (1972c; 1973, 220–31); see also Lefebvre (1972, esp. 99–103); and extended discussion of city origins in Bird (1977a, 27–46).

While scientific Marxism provides the precision to match the qualities that Harvey had earlier extolled in the deductive-nomological approach, it is ever-constraining in basing everything on the primacy of the material foundation of the mode of production. He has more recently suggested that elements in the super-structure of society may also have a vital input to bring about change. This is proof that Harvey has turned into a critical Marxist, to use Gouldner's (1980) classification, or exhibits what scientific Marxists would call 'revisionism'. This was foreshadowed in the suggestively titled 'Monument and myth' (Harvey, 1979), which showed how the Basilica of Sacré-Cœur has symbolized different things to different classes throughout history. We are shown the play of contrasted consciousnesses over and within the material presence of the edifice. Notice that in the following 1981 passages, Harvey starts in scientific Marxist manner, and then the italicized *buts* indicate the beginning of critical Marxism:

The life of mind is an integral part of social existence and is thereby tied to the mode of production of material life—science develops as a productive force, different social classes (defined in relation to production) develop distinctly different cultures. *But* at the same time, with the proliferation of the division of labour and the increased production of surpluses, intellectual activity develops a certain autonomy. Consciousness can begin to 'flatter itself that it is something other than consciousness of existing practice'; culture evolves a life of its own.

. . . the cultural geography of capitalism cannot be understood without referring to the various technological, economic and political structures of this mode of production or its class character—to deny this is idealism. *But* such a geography cannot be written with reference to these structures alone. Autonomous aspects have to be taken into account as well. (Harvey, 1981, 211; my italic)

Strangely enough, a passage-at-arms between Wagstaff and Day (1980) reveals that Engels himself changed from a scientific Marxist interpretation (Engels, 1892, originally 1877, quoted in Wagstaff 1980, 148) to a more critical Marxist stance where some autonomy was allowed to elements of the superstructure (a letter from Engels, 1910, quoted in Day, 1980, 145). The move to a critical Marxism is again symbolized in the very title of Harvey's 1985 book, *Consciousness and the urban experience*:

The urbanization of capital is an objectification in the landscape of that intersection between the productive force of capital investment and the social relations required to produce an increasingly urbanized capitalism. But this implies that we should look also at the implications for political processes. The 'urbanization of consciousness' has, I therefore submit, to be taken as a real social, cultural and political phenomenon in its own right. (1985, p. xviii)

We appear to have moved on from a tight Marxist structure, a 'real foundation, on which rises a legal and political superstructure and to which correspond definite forms of social consciousness' (Marx, 1962, originally 1859).

An 'alternative [to Paterson's 1984] assessment' of Harvey's published work can be found in Johnston (1986*b*) who recognizes him as a recruit for 'realism' (see below in this chapter), though Harvey (1987, 373) has since criticized this approach (see also below). He has said: 'I believe the claim of Marxian analysis to provide the surest guide to the construction of radical theory and radicalizing processes still stands'; and that 'the Marxist approach for me is so much part of my being now that I can't think in any other terms' (Harvey, interviewed in Peake and Jackson, 1988, 19), describing himself as a 'restless analyst' (1985, p. xi); but he believes the following, which would cause some Marxist eyebrows to rise:

. . . there are good grounds for thinking Marxist formulations incomplete, still open-ended, and by no means omnipotent in confronting the realities of daily life. (Harvey, 1988)

and

the thing I like about it [Marxism] is in some ways how extraordinarily open it is to reformulation, rethinking. A lot of people associate Marxism with dogmatism . . . My experience of it has been a constant dialoguing, dialectic if you will, of interrogation of conceptual apparatus, thinking through problems and thinking about politics and it's extremely open and fluid in that way. (Harvey, interviewed in Peake and Jackson, 1988, 19–20)

And so, recalling all the structures associated with most forms of Marxism, perhaps we might do better to call him the 'restive synthesist'.

Interlude: national schools, with special reference to Soviet geography and a battle of the -isms

This two-chapter parade of -isms, like the rest of the book, is heavily dependent on Anglo-American sources, to the neglect of other national schools, like compartments in the great edifice of world geography, the walls of which are different languages. However, at the end of a most useful international survey (Johnston and Claval (eds.), 1984), Claval believes we now have a simple system: 'one-way communication from the English-speaking countries to the rest of the world' (p. 285). Another brief international survey consisting of the highlights of the decades

1900–80 is to be found in Buttimer (1983, appendix B, 261–74). Besides language, lack of research funds confines research spatially, promoting 'national parochialism', for government agencies are more likely to provide funds for 'relevant' research at home. Understandable as this might be, we are here witnessing a propinquity-led development rather than an intellect-led quest. For example, you can bet that British industrial and social geographers are almost all now studying problems in Britain.

One national school of particular interest on at least four counts is the Russian school, because of its long history, with a great richness stretching back long before the revolution (Hooson, 1984, 82–6); the pronouncement on geography by Stalin; the Anuchin affair (see below); and, particularly relevant in the context of this chapter, the relationship between Soviet geography and radical geography in the West. Hooson has perhaps been the principal ambassador for Russian geography since his first paper in 1959, though Shabad's editing of *Soviet Geography*: *Review and Translation* since 1960 has made the fruits of the Russian school widely available.

In the 1930s it was the Russian view that Western geographers firmly believed in physical environmental determinism, and this was contrary to Marxist-Leninist principles. No less an authority than Stalin himself felt it necessary to lay down the following principles for geography in his September 1938 pronouncement:

... the geographical environment indisputably is one of the constant and necessary conditions of society and, of course, influences the development of society; it accelerates or retards the speed of development of society. However, its influence is not a *determining* influence, inasmuch as the changes and development of society proceed incomparably faster than the changes and development of the geographical environment ... Geographical environment cannot be the chief cause of development of that which undergoes fundamental changes in the course of a few hundred years.

Stalin added that the force determining the material life of society, the character of the social system, and its development is the

method of procuring the means of life necessary for human existence, the mode of production of material values—food, clothing, footwear, houses, fuel, instruments of production, etc.,—which are indispensable for the life and development of society. (Stalin, 1938, quoted by Matley, 1966, 102)

The result of this fiat was a complete split in Russian geography between an overwhelmingly dominant physical geography and an emasculated economic geography, a state of affairs which lasted for twenty-six years. It was the Anuchin affair that set changes in motion. In 1960 Anuchin (1977) argued for a unified approach in geography and against the dichotomy of physical and economic geography, with supposedly separate 'laws' in each field. There were many sensational confrontations between the Anuchin party and the guardians of the old order, a conflict 'comparable in Soviet terms with the more notorious Lysenko controversy in genetics and biology in general' (Hooson, 1984, 91). This 'battle of -isms' included Anuchin's public dissertation defence at Moscow University in 1962, attended by hundreds of people, when he failed to obtain the necessary two-thirds majority for success. The battle was finally over in 1964, significantly the last year of the more open Khrushchev regime, when

a leading ideological spokesman, L. F. Ilyichev (1964), in a statement before the Presidium of the Academy of Sciences, denounced the Stalinist definition of the environment as 'a purely natural category'— laid down 30 years before, and the fact that this edict had become the pretext for the construction of 'an insurmountable wall' between nature and society, with deleterious effects on the Soviet economy and planning processes. (Hooson, 1984, 92–3)

Finally, to conclude this mainly Russian entr'acte, it is interesting to read a Russian review (Lavrov et al., 1979; 1980, in English) of Peet's (1977) edited volume, *Radical Geography*. The Russian authors noted the Western radical geographers' sharp critique of bourgeois geography's techniques, including quantitative geography:

The critical side of the radicals' approach to various problems is expressed far more sharply than any positive aspect. Some of the criticism tends to be overblown, proclaiming, for example, as reactionary even those problems and methods that would appear to be of some significance in socio-economic geography (the development of mathematical techniques, work in behavioral geography, etc.). A striving to sociologize geography thus goes over to the extreme of rejecting the object of study specific to geography and questioning the existence of geography itself. [At this point Slater, 1977 is quoted; a similar point of view has also been forcefully expressed more recently by another radical, Eliot Hurst, 1985] . . .

The excesses of radical geography derive to a large extent from the fuzziness and intricacy of its theoretical base, which is only now beginning to be formulated (and in a great variety of ways). As of now it is an eclective conglomerate, on the one hand, of Marxist propositions and, on the other hand, of anarchist and ultraleft views. These excesses become understandable if one considers that most of the theorists of radical geography (judging from the book under review) have so far achieved understanding of only a few Marxist-Leninist propositions even though some of them proclaim Marxism-Leninism. (p. 319)

The review went on to complain that the Westerners were unfamiliar with construction, territorial planning, and regionalization in socialist countries, and that they quoted not one reference from geographers in those countries.

So Western radical geography, though overwhelmingly Marxist, appears very different from Soviet geography, itself proclaiming that 'the only scientific and promising path for science is one oriented to Marxist-Leninist theory' (Lavrov et al., 1979; 1980, in English, 320).

Loosening the structure: anarchism, transactional-constructivism, symbolic interactionism, structuration, transcendental realism

We have already noted how the tight structural control within scientific Marxism has been very slightly relaxed in what has been called 'critical' or 'humanistic' Marxism. There have been other moves to avoid tightness of control. Peet (1977) flirted with anarchism because, while increased central state (or structural) control over investment in social resources would be a move towards an egalitarian society, he saw a problem: 'The problem with this model, however, is bureaucratization, with a lack of sense of control over one's environment. An attractive alternative model, developed in its most sophisticated form by the anarchists, involves decentralised, worker ownership of the means of production and a linked system of community control over environment' (p. 121). This solution is condemned by Harvey (1973) in his scientific Marxist phase:

One way to rectify this situation is to negotiate with neighbouring communities, but the problems of non-centralized information gathering and negotiation costs (including those to be imputed to a delay in decision making) are likely to make this an inefficient way of rationalizing the provision of these services . . . Locally financed government is a disastrous proposition—it will simply result in the poor controlling their own poverty while the rich grow more affluent from the fruits of their riches. (p. 93; see also Harvey, 1984, 9)

The next three -isms in this section are borrowings from sociology which have been used by geographers to avoid the extreme tightness of a structural approach (as in scientific Marxism) without the unbridled voluntarism based on the values, attitudes, and actions of individuals (see Table 15). We have here the dilemma of the advantages and disadvantages, often complementary, of two polarities, not unlike the contrasts inherent in different scales of approach. One proposed solution is the dialectical method, often referred to as 'the negation of the negation'. This phrase comes from Hegel, who inspired Marx to the method of dialectical materialism: 'The negation of the negation—the working class is defined in terms of the non-ownership of the means of production, it has as its goal the abolition of all property and it therefore dissolves itself into all other classes, in the revolutionary process' (Burgess, 1976, 39 n. 25). William Blake, a contemporary of Hegel, put the negation of the negation quite simply: 'You never know what is enough unless you know what is more than enough,' (Elster, 1986, 34, quoting Blake). One must experience extremes alternately in order to make a judgement, and this alternation is the dialectic. This method is found in the social theory of Berger and Luckmann (1966) as here summarized by D. Gregory (1981c, Table 1, 11): 'society forms the individuals who create society in a continuous dialectic: society is an externalization of man, and man a conscious appropriation of society'. There is a form of the dialectic method in the transactional-constructivist position, as summarized by Moore and Golledge (1976, 14); and I understand that in their following application of it to man and environment, 'individual' could be substituted for 'organism' and 'society' for 'environment': 'Transactions between the organism and the environment are viewed as mediated by knowledge or cognitive representations of the environment; but these representations are treated as constructed by an active organism through an interaction between inner organismic factors and external situational factors in the context of particular organism-in-environment transactions.' This perspective has been carried over into a general textbook on behavioural geography by Walmsley and Lewis

TABLE 15.   Twin dangers of scientific Marxism and a hyper-individualistic action theory[a]

	Scientific Marxism[b]	Hyper-individualistic action theory
Basis	Structural historical materialism	Neo-classical economic and behavioural traditions
Assumptions	Social structures being driven towards some pre-ordained end; capitalist mode of production as autonomous, self-determining structure	Social explanation based on individual values, attitudes, desires, and actions; individuals have free choice
Constitution of society	Society is a reality constraining upon human agency	Society, classless and harmonious, is constituted by intentional action
Dangers		
	Individuals seen as passive agents, prisoners of fate because evolution is at level of structures	Social origins of individuals' preferences ignored; ignores roles of State and institutions (economic and social)
	Reification of structures	Voluntarism[c] of individuals at expense of their context

Note: The table attempts to illustrate general debate in the social sciences by juxtaposing extremes; cf. the following: 'The recent debate in the philosophy of the social sciences has turned on two related polarities, that between a 'subjectivist' and an 'objectivist' pole, and that concerning the relationship of agency to structure. The first has been haunted by the specter of philosophical idealism, the second by that of a world without agents' (Manicas, 1987, 267).

[a] In social theory the two approaches are typified respectively by Emile Durkheim (minus the Marxism) and Max Weber (minus the extremism).

[b] This term is taken from Gouldner (1980), and is in contrast to critical Marxism (or humanistic Marxism, or Neo-Marxism), which does have a few voluntaristic elements, but nothing like the reversal of emphasis in an individualistic action theory.

[c] See an interesting diagram of the relative strengths of determinism and voluntarism of various writers, according to subject area, within Marxism, and within the structurationist school, Thrift (1983, Fig. 1, 27).

Sources: Based on D. Gregory (1981c, Table 1, 11) and J. S. Duncan (1985, 176–7).

(1984, 45). Another use of the dialectic method is exhibited by Duncan (1978), via the concept of symbolic interactionism, to aid the analysis of the tourist's cognition of the environment, though he seems later to have abandoned this approach in favour of structuration.

In moving on to structuration we proceed from continual dialectic dualisms to the continuous dialectic of fused polarities (Table 16). In this formulation the societal structure and the

TABLE 16. The dialectics of structuralism

| Dialectic (contradictions in thought-systems) | Dualism $x — \blacktriangleright y$ and/or $x \blacktriangleleft— y$ | Polarities treated as both opposite and distinct (relationships between opposites *such as individual and society*, presented in the form of a one-way directional arrow) = deconstruction and partial recombination of what already exists |
| | Duality $x \blacktriangleleft\!\!\top\!\!\blacktriangleright y$ xy | Interdependence between polarities is equal in weight (relationships between opposites *such as purposive action and society* are of equal weight; each exerts a determining influence on the other) = creating something new |

Source: based on text in Gregson (1986, 185 and 201).

individual create each other by, respectively, the outcomes and actions of their existence. If that seems opaque, here is a summary by the founder of the concept:

The concept of structuration involves that of the *duality of structure*, which relates to *the fundamental recursive character of social life and expresses the mutual dependence of structure and agency*. By the duality of structure I mean that the structural properties of social systems are both the medium and the outcome of the practices that constitute those systems ... (Giddens, 1979, 69; for four diagrammatic models of structuration following a theoretical exegesis (no critique) see Moos and Dear, 1986)

Two points about structuration need to be stressed: first, it is of particular interest to geographers because of the inclusion of a spatial ingredient:

The structural outcomes of social practices are premised upon their time–space characteristics, since these time- and space-defined practices serve simultaneously as patterns of interaction and, through their impacts on human experience, as foundations for the motivation to future practices. Power, which is the glue that holds institutions or structures together, is built from particular forms of time–space distanciation of social practice, because they make possible specific forms of control over allocative or authoritative social resources. (Storper, 1985, 408)

Second, Storper later points out that structuration, with its inclusion of practical knowledge as both producing interactions and produced by them, is different from functionalism where: 'the evolution of an institution, system, or behavioral pattern is *unintended* and *unrecognized* by the members of the beneficiary group, but is maintained by a causal feedback loop passing through the beneficiaries' (Storper, 1985, 410–11, with Elster's (1979, 28) definition of functionalism in sociology quoted as a footnote). Anthony Giddens, the begetter of structuration, is a Cambridge sociologist, and Gregory has been the principal exponent of structuration for geographers (for a summary see D. Gregory, 1986*b*), though Pred (1984) and Duncan (1985) have also been ambassadors. Perhaps the most accessible summary is to be found in two pages by Johnston (1983*b*, 103–5), and the following is an extract from his last paragraph on the subject:

[Giddens] identified three concepts regarding society: first, the *system*, or set of regular social practices; second, the *structure*, the (changing) rules and resources which guide the system; and third, *structuration*, the conditions whereby members of systems govern the continuity and/or transformations of structures, and thereby influence the reproduction of the systems themselves. What Giddens is proposing, therefore, is the development of a mode of analysis which pays substantial attention to the role of human agency in the transformation of structures . . . Structuration accepts part at least of the Marxist arguments regarding the role of the infrastructure, but argues for the important role of human agency in both the realization of the structural processes (i.e. the observed world of the empirical researcher) and the modification of the processes themselves.

The basic attraction of structuration is that processes are seen as emanating 'upwards' from the individual and 'downwards' from the structure, occurring simultaneously, and affecting each other and the system in which they are embedded even as they happen. Among many other sources, Giddens (1984, 116) has acknowledged his debt to time geography, and because structuration involves individuals and societies reflexively influencing each other in specific time–place contexts, it is little wonder that geographers have been interested in the links between the concept and human geography (e.g. Thrift, 1983).

And thus we arrive at the final -ism in my parade, transcendental realism (Fig. 13), but remember that geography is an open system and the 'finality of this finality' is but temporary. E. Graham (1988) has carefully distinguished metaphysical realism from transcendental realism. The first states that the world is as it is independent of human understanding of it and that empirical verification can be made by the correspondence of theory to the real world (the correspondence theory of truth). Transcendental realists, however, use terms like 'practical adequacy' for corroboration, and believe that phenomena in the real world are structured and possess causal powers and liabilities. Graham therefore concludes that transcendental realism is very different from metaphysical realism. It can also be distinguished from an empirical realist view, where regularities are discovered on the basis of observation. Sayer (1988) confirms that transcendental realism holds that there are causal powers embedded in the social objects of study and that social phenomena are intrinsically meaningful. Meanwhile, Cochrane (1987) had emphasized the importance of transcendental realism for geography within the social sciences.

The distinction between necessary and contingent relations which is so important to [transcendental] realism has been presented as a means of acknowledging the uniqueness of different places, without giving up the idea that their development also reflects the interaction of general processes. As a result geography can no longer be seen as the Cinderella subject of the social sciences, borrowing most of its methods and ideas from other disciplines. Instead it is able to make its own claims that social, economic and political processes cannot be discussed without being informed by geographical analysis. (Cochrane, 1987, 354; this is

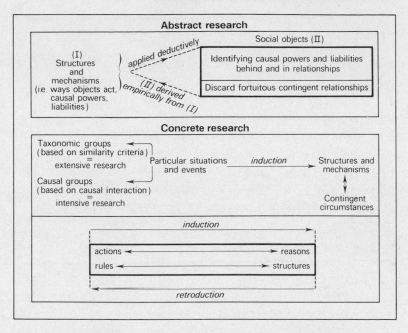

FIG. 13. *Transcendental realism: its abstract and concrete (extensive and intensive) types of research, their internal constituents (including structures), and three methodologies: deduction, induction, and retroduction*

Note: The three methodologies are shown in italics. While this diagram attempts explanation (*erklären*) via an analysis of transcendental realism, many who advocate this approach might argue that the diagram is itself an example of positivist dichotomous thinking, and transcendental realism demands sympathetic understanding (*verstehen*) in which all its included strategies are seen as synchronic complements of each other.

Source: Derived mainly from the text in Sarre (1987, 4), with additional help from Sayer (1988, personal communication), though neither bears any responsibility for this schematic actualization.

part of Cochrane's presentation of what he calls the 'new geography' and he goes on to present criticisms and then comments on the criticisms, see below.)

A genealogy of the introduction of transcendental realism into geography (Table 17) shows that there have been attempts to join it to structuration but allowing actors (agents, decision-makers) in their various spatial locales to interpret the structures in which they recognize themselves as being enmeshed. Structures are seen

TABLE 17.   Outline history of transcendental realism in geography

1975	Source of transcendental realism Bhaskar (1975)[a]
1979	Applied in industrial geography Massey and Meegan (1979)[b]
1981	Relationship of transcendental realism[c] to other 'realisms' Gibson (1981)
1982	Applied in industrial geography Sayer (1982a)
	Applied to economic geography Sayer (1982b)
	Related to other philosophies of science Keat and Urry (1982)
1983	Applied to social geography Allen (1983)
1984	Applied to social science generally Sayer (1984)
1985	In the analysis of space and time Sayer (1985b), Urry (1985)
	Allied to structuration Soja (1985)
1986	Summary to date D. Gregory (1986a)
1987	Discussed as a method Allen (1987): review of Sayer (1984)
	Applied to social geography Sarre (1987)
	Criticized as a methodology Archer (1987, 390–2); Harvey (1987, 373)
	Acknowledged as a 'new geography', with two principal criticisms calling it 'partial' and attempting to develop a critique of it as the 'new structuralism of locality[d] (Cochrane, 1987)
1988	Symposium on Geography and Realism[e] Papers included Sayer (1988), Allen (1988), Livingstone and Graham (1988); see also E. Graham (1988)

[a] 'It regards the objects of knowledge as the structures and mechanisms that generate phenomena . . . These objects are neither phenomena (empiricism) nor human constructs imposed upon the phenomena (idealism), but real structures which endure and operate independently of our knowledge, our experience and the conditions which allow us access to them. . . . That generative mechanisms must exist and sometimes act independently of men and that they must be irreducible to the patterns of events they generate is presupposed by the intelligibility of experimental activity' (Bhaskar, 1975, 25 and 52).

[b] Realism is not explicitly mentioned, and Bhaskar (1975) is not referenced therein, but see opening discussion (pp. 159–64).

[c] The nearest of Gibson's nine realisms to transcendental realism appears to be his 'scientific realism' (1981, 157–9).

[d] 'Localities' are seen as more microscale than conventional regions by Jonas (1988; 'small towns, rural parishes, working class communities' are his examples, p. 101); '. . .

structures might transcend a variety of spatial scales and many localities. Such a method would have to put people before place. To that end, any appeals for a new regional geography of localities would seem misplaced' (p. 108).

^e Arranged at the Institute of British Geographers' Annual Conference, Loughborough, 7 Jan., by the History and Philosophy of Geography Study Group.

as within this scheme rather than forming a determining framework, although there is debate as to the connections with Marxism. As E. Graham (1988) has pointed out, two steps would be necessary to go from realism to Marxism,

the first involving the move from the metaphysical realist claim to the transcendental realist view that the world is stratified such that structures operate as generating mechanisms for events; the second step being the move from this general picture of structures as generating mechanisms to the particular claims derived from Marxist theory about what the most important of these structures are.

On the one hand, the important embedded structures could be Marxist in type; on the other hand, Sarre (1987, 6) complains that in Sayer's (1984) version of transcendental realism there is no discussion 'in detail [of] how realist knowledge can change reality'. But the epistemology does have room for idiographical variations (contingent relations)—a method of putting space back into social science. And so Massey in her empirical demonstrations can claim that 'Geography matters' (1987), with 'geography' here in the popular usage of local environmental conditions: physical, socio-economic, and of course structural: '*we can only understand social processes . . . if we also understand something of their geography.* While the first argument basically said, "in order to understand space you need to understand society", this argument says "yes", but to understand society you also need to understand space' (p. 6). In transcendental realism, as I understand it, capitalists and other actors are allowed to interpret the structural components of their socio-economic context, and alter their strategy accordingly, not only deterministically according to Marxian mechanisms, but also in ways through which they can work with those mechanisms, or even circumvent them, to achieve what they think will deliver the greatest advantage (profit) to themselves. In the case of firms from the smallest to the multinationals this leads to the patchy over-accumulation of capital and to increased spatial inequality.

There seems to be a clear progression: scientific Marxism, critical Marxism, structuration, transcendental realism, with each allowing a progressively diminishing constraining role for structures. The idea of everything happening at once to all elements connected to each other is not easy to explain in sentences where meaning unrolls serially; and this may account for the fact that, where geographers have tried to sell structuration and transcendental realism, they are scarcely an easy read. When an attempt is made actually to conflate both approaches, the following results:

In its attempt to avoid the limitations and distortions of positivism and idealism, its affirmation of a redefined synthesis of structuralism and hermeneutics, its situation of social theory and social practice in the conjunctural effects of time and space, its adaptation of an essentially Marxist notion of praxis while simultaneously subjecting Marxism to vigorous reconstructive critique and 'modernisation': in all these, the new theoretical realism directly adjoins with and helps to reinforce the connections between the structurationist school and the contemporary transformative retheorisation of spatiality. (Soja, 1985, 123)

This is just one sentence! You can eventually see what is meant, but it is quite a sentence to have to do so.

Strictures on structures

Stricture ... An incidental remark or comment; now always an adverse criticism.

Oxford English Dictionary, meaning 5.)

A structuralist approach asserts that there actually exist structures behind the world of appearances. By definition, we cannot see these structures, so we have to deduce them, and to do so requires sounder bases than 'there must be a better socio-economic system than capitalism' or 'everything is connected structurally to everything else'. It is as easy to fall into the trap of the reification of structures as to commit the ecological fallacy whereby individual characteristics are derived from the aggregate characteristics of a population (scale reductionism). All forms of Marxism are basically optimistic about the perfectibility of human nature and human society and share with Christianity a belief in a better life in the next world—which would be that society supported by the mode of production that succeeds capitalism. As far as geography is concerned, little can be derived from Marx for physical geography; the distribution and process-systems of natural resources and their conservation were hardly nineteenth-century preoccupations. This does not mean that physical geography is immune from current Marxist attacks. Western-style capitalism has 'destructive commodity attitudes' towards environments (see Table 13), and physical geographers are dubbed technocratic supporters of the status quo, depending for research support on the continuous health of capitalist states, which they thereby implicitly support rather than help to overthrow. Incidentally, Marx had also little to say on racial or ethnic problems, nationalism, and the dangers of global conflict, all part of a modern geographical agenda.

Not many non-Marxist geographers publish counter-arguments to historical materialism in geography. This is strange because it is a sure way of getting published works into citation indexes, for Marxists are very litigious and always counter-attack. Cole (1986) is by far the longest sustained polemic against the spread of historical materialism into human geography, but there are others: Ley, 1978, 46–9; Clark and Dear, 1978: Wagstaff, 1979; Walmsley and Sorenson, 1980; Duncan and Ley, 1982; and Corbridge (1986). The last is a book, with the subtitle 'a critique of radical development geography', which builds up to a criticism of a static structural approach to what are the problems of a constantly changing world affected by a range of agents causing the changes.

Amongst these agents I would include the environment, the transnational corporations and the nation-state; the point being that an account of the changing dynamics of the modern world system must have recourse to a range of 'explanatory variables' which occupies a middle-level between capitalism-in-general and the individual and his or her class . . . our accounts of differential development must recognise that the dynamics of a changing capitalist world economy are always mediated by conditions of existence (population growth rates, gender relations, state policies and so on) which vary in space and time and which are not directly at the beck and call of a grand 'world system'. (pp. 246–7)

A paper by Saunders and Williams (1986) on the relationships between the 'new conservatism' and the recent and future developments in urban studies not surprisingly prompted a 'spirited counteroffensive', 'an aggressive counterclaim' (the descriptions are by Dear, 1987, 363 and 364) from Harvey (1987), and no less than nine other comments in a 67-page debate, preceded by an editorial summary (Dear, 1987). This debate ranged widely, including topics such as: the retreat from scientific Marxism ('Althusserianism'); the question of whether or not transcendental realism is a viable alternative to historical materialism; the many 'silences' in Marxism 'about things labelled as urban and regional problems' (Ball, 1987, 393) in 'a post-Fordist era' (Storper, 1987, 420); and discussion of post-modernism (Thrift, 1987, 404; Storper, 1987; postmodernism and modernism are referred to below), with scientific Marxism placed squarely in an outmoded modernist perspective.

The monograph by Cole (1986) deals with contributions by Marxist geographers on classes in Western societies (social geography), stages of growth (developmentalism), and capitalism (social and economic geography). Class categories in Western societies are now so complex that, with certain obvious exceptions, it is difficult to see who is exploiting whom in the overaccumulation of capital. Employees are obviously working class, but they may invest surplus income and inherited wealth in shareowning, so that presumably they metamorphose into capitalists when they retire. The shrinkage of the abject poor through welfare programmes and the emasculation of great wealth by taxation have been accompanied by the great rise of the 'middle' income ranges (see Bird and Witherick, 1986, Fig. 1, 307). But mention of welfare only excites comments from Marxist geographers that such programmes are merely attempts by the capitalist system to legitimize itself by anaesthetizing potential opposition groups. The system must be changed into one that does not merely confront present inequalities but is based on the principle of communistic equality as a ruling principle. Popper had a comment on this:

For nothing could be better than living a modest, simple and free life in an egalitarian society. It took me some time before I recognized this as no more than a beautiful dream; that freedom is more important than equality; that the attempt to realize equality endangers freedom; and that, if freedom is lost, there will not even be equality among the unfree. (Popper, 1976a, 36)

I have to point out that a mention of Popper in an argument with Marxists is called by them the 'vampire trick': attempting to get the 'same effect on marxists as brandishing a cross in the face of the vampire in Dracula' (Duncan and Sayer, 1980, 195).

A framework involving stages of development is of course not particular to Marxism. Cole (1986, 54–5) points out that Vico (Bergin and Fisch, 1961), Spengler (1926), and Rostow (1960) have promoted developmental schemata; to which we may add North (1955), Bruton (1960), a review by Hoselitz (1960), and the time-based scheme of Kondratieff, which has been combined with space in a 'space–time matrix' by Taylor (1985, 10–21). The peculiarity of Marx's historical materialism is that Western-style capitalism is definitely posited as the penultimate stage, the accelerating crises of which usher in the millennial communistic final stage. This puts Marxists in a dilemma. If they expose these crises to public attention, they may be delaying the millennium they fundamentally desire. Remember Harvey (1981, 209): 'The aim is not merely to understand the world but to change it.'

Geography, in common with Western capitalist society, is made up of countless human endeavours. Any human endeavour entails both gains and losses; one obvious loss is the opportunity cost of the expenditure of time. So that any advantage gained by human endeavour must be measured in terms of *net* gains. As a result every human endeavour can be criticized. The difficulty with historical materialism, and the radical geography based upon it, is that it does not, or cannot, admit that its advocated society emplaced by revolution will have any disadvantages. And therein lies a fatal flaw. It is true that as one gets older, one begins to hear oneself saying more frequently: 'I believe in change but I have to be sure that the change is better than what I have now.' This is not the remark of a geriatric reactionary but comes with the wry experience that life contains simultaneous ups and downs, goods and bads.

The above paragraph would cut no ice with Peet's (1978) reply to Clark and Dear's (1978) claim that radical geography had made little impact on geography:

Revolution . . . means a change in the nature of society and in human nature itself (p. 362) . . . The problem with joining the mainstream (even as its left current) is that marxist analysis is deliberately fragmented—pieces of it are plucked out, laundered of revolutionary content and synthesized with the existing ideological structure. And the 'radical' geographers who participate are integrated into the established power structure, rewarded by it, and rendered ineffective as critical critics. (p. 364)

This is clear enough: no compromises, and 'a call for revolution is the only possible outcome of radical research' (p. 363).

A historical materialist would argue that if the capitalist system develops a crisis, predictions are confirmed. If, on the other hand, the system continues to succeed, this is a confirmation of the tendency to over-accumulate capital; and if the workers under capitalism achieve material gains, here is merely the system's act of legitimation, buying off the workers to avoid erosion of their complicit political support. The basic thesis of the evils of capitalism held by historical materialists is therefore irrefutable, or as Kirby and Pinch (1983, 241) put it in their comment on the contribution of structuralism to the problems of territorial justice and service allocation, 'like some epistemological game of football, any results—win, draw, or lose—can be accommodated into the expectations'. But historical materialism posits a fundamental world-wide relationship energizing societies—the mode of production, or, in geographical terms, the transformation of resources, and this is the reason for its strong currents within social sciences including geography. Another four would-be ruling frameworks are Christianity, Islam, science, and capitalism, although the last two have no one source or bible of coded beliefs. Cole has produced a table showing how four of these 'authorities' view themselves and each other (Table 18).

The four world-views are so comprehensive that it is to be expected that they are largely antipathetic to each other, but some geographers believe that a compromise between Marxism and humanism is feasible: 'Advocacy of a move by historical geographers towards the adoption of a Marxian humanism is not an extremist view. It is instead a plea for balance, not only philosophically but also methodologically and technically' (Baker, 1981, 243). Here is a *non sequitur* of breathtaking naïvety. By definition, Marxism cannot compromise with another system. As Eyles (1981, 1375–6) points out, to borrow attractive bits of Marxism to improve analysis and explanation strikes at the heart of its dialectical method and holistic claims. The holism also would deny separate identities for spatial processes and for geography

TABLE 18. Authorities on authorities

This on this	Christianity (Bible)	Science	Capitalism	Marxism
Christianity → (Bible)	The word	Incomplete: God in the gaps	In moderation	Potential heresy
Science	No miracles	On towards the truth	Source of research funds	Source of research funds
Capitalism	Should not get in the way	Useful applications	The best economic system	Highly undesirable
Marxism	The opiate of the people	Useful applications	Terminally ill	The ultimate society

Source: Cole (1986, Table 7.3, opp. p. 80) adds this comment: 'Of the four major authorities . . . Science seems to be the most successful in preventing the emergence of alternative authorities. Christianity has become terribly fragmented over nearly 2000 years. Marxism has different approaches and different faces even where officially accepted and applied, as in such diverse environments as the USSR, China, Hungary, Yugoslavia, and Cuba. A market economy is predominant in most countries of the world but alongside varying degrees of state intervention or cooperative ownership of means of production' (p. 115). Islam and other possible 'world authorities' are omitted, no doubt to avoid overcomplicating the table.

(p. 1377), and leads on further: to deny the existence of separate social sciences such as geography because all disciplinary boundaries are counter-revolutionary (p. 1378; see also Eliot Hurst, 1980). 'In total . . . radical reconstruction cannot be carried out without a revolution. It is not geography as currently or potentially practiced that must be buried, but also the bourgeois academy and the oppressive socio-economic system of which it is but one part' (Eliot Hurst, 1985, 85). This must qualify as one of the most extreme and uncompromising statements about the subject ever made by a professional geographer.

Despite his own warnings, Eyles puts forward a framework which tries 'to incorporate and transcend Marxism', not realizing that Marxism becomes a poison pill when combined with other epistemologies—not that this has prevented many attempts (combination with psychoanalysis, Fromm; with phenomenology, Merleau-Ponty; with existentialism, Sartre; with structuralism, Althusser). Often Marxists struggle with a problem deep inside their philosophy, neatly demonstrated by Eyles (1981, 1373):

If the economy always determines in the final instance, the other levels [i.e. legal and political superstructure and definite forms of consciousness] become mere epiphenomena of the economy . . . If they are not [dominant], is one simply presented with a causal model of structural determination which denies a role for man in the making of his own history and which suggests his impotence in effecting change? Indeed, these problems have led some Marxists [e.g. Althusser and Balibar, 1970, 100, and the later Harvey, see 1981, 211] to speak of the relative autonomy of levels, a truly Comtean exercise which does not seem particularly Marxist, differing little from the multicausal perspectives of Weberian or even empiricist sociological analyses.

A natural postscript to the account of Marxism and geography is a reference to an attempt to promote a 'post-Marxism', based on actions by a 'plurality of social agents', not just the proletariat, and including abandonment of the idea of a communistic millennium:

. . . we have indicated that the political transformations which will eventually enable us to transcend capitalist society are founded on the plurality of social agents and of their struggles. Thus the field of social conflict is extended, rather than being concentrated in a 'privileged agent' of socialist change. This also means that the extension and radicalization of democratic struggles does not have a final point of arrival in the achievement of a fully liberated society. (Laclau and Mouffe, 1987,

106: a reply to Geras, 1987, itself a critical review of Laclau and Mouffe, 1985)

Structuration is beginning to attract critical comment in the geographical literature (Storper, 1985; Gregson, 1986, 1987). It is as if Giddens, having seen the pitfalls of earlier sociological dualisms, either too deterministic or too voluntaristic, then bolted together a duality which would withstand criticisms that former frameworks had attracted. Nevertheless, a major difficulty arises with the degree and kind of agents' intentionality: 'The identification of structure with constraint is also rejected: structure is both enabling and constraining' (Giddens, 1979, 69). Is it? Who says so? Giddens says so. That is the point. Structure is 'seen' by agents, and academic commentators, as enabling or constraining. But then we need to ask upon what basis do they see and act. Giddens would reply that the supraindividual social world impinges on the individual by means of the latter's non-discursive (i.e. 'not able to be fully articulated') practical knowledge.

I want to argue here that Giddens does not actually specify how human agency can be based on nondiscursive practical knowledge. Practical knowledge is itself a *product* of structured interaction in time and space. It is unclear how it can both produce interaction and be produced by it. As things now stand, his action theory is potentially functionalist itself. (Storper, 1985, 410)

Gregson (1986) goes even further in criticizing structuration. She accuses Giddens of 'fairly superficial' use of Hägerstrand's time geography (p. 193), which began as a rather deterministic system (via constraints), but which, following humanistic criticism, has metamorphosed through Hägerstrand's later work into something more voluntaristic (stressing contexts). Giddens refers only to the first of these formulations. A further criticism from Gregson is the difficulty of employing or exemplifying structuration in empirical research because of the very stress on the duality approach. She quotes from one who has tried: 'whilst theoretically attractive, the simultaneous and mutual "causal" influences of agents on society and society on agents makes it difficult to apply this vision to concrete instances of social activities; it is difficult to know how one is to cut into the data' (C. Smith, 1983, 3). Gregson's conclusion from a survey of the very few empirical studies so far to use a structuration perspective is that empirical research is able to provide insights into 'either agency (human practices) or institutions but not both together' (p. 199; cf. transcendental realism, Fig. 13) and that the focus is easiest to apply at the microscale. This is because, in the Giddens scheme, structure has no existence until some form of human action takes place. Kellerman (1987) points out that when structuration is advocated at a particular scale, say the city level as in Giddens (1985), something may be lost at other levels, regional, megalopolitan, national, and international. Apparently, we cannot have it all ways: 'Focusing on one specific era (synchronic analysis) would misleadingly present time as static. Using the more "dynamic" diachronic approach would detract from an integrative analysis of a more specific static period' (Kellerman, 1987, 273).

Gregory (1988, and 1992, in press) is currently debating the relationship between human geography and social theory, and, in particular, the tension between modernism (a late nineteenth-century and early twentieth-century movement, last heir of the Age of Reason) and post-modernism (a reaction to modernism; for particular reference to geography see Dear, 1986; Soja, 1987; Dear, 1988; Dear et al., 1988; and Storper, 1987, who uses the term 'post-Enlightenment'). Under the admittedly ambivalent label of post-modernism there are tendencies, notably in the humanities and social sciences, for something like an 'anything goes' approach wherein concepts of order, procedural rigour, rationalism, or indeed the very idea of an academic 'discipline' are dubbed as examples of an outmoded modernism. The alert reader will recall Feyerabend (1975) and his complete title: *Against method: outline of an anarchistic theory of knowledge*. There appears to be some doubt as to whether modernism is believed to be over and done with or whether post-modernism represents its radical reworking (Baynes et al., 1987; Lyotard, 1987).

For the past decade the term 'postmodern' has been used in so many different ways and with so many different senses as to render precise specification of its meaning impossible. . . . The contours of this shift seem clearest in literature and the arts, where postmodernism contrasts with the aesthetics of classical modernism. To take the clearest case, in architecture the repudiation of the glass and steel functionalism of the International Style [e.g. Le Corbusier and Mies van der Rohe] typically takes the form of a play of historical allusion, eclectic pastiche, appropriation of local traditions, and a return to ornamentation and decoration [e.g. James Stirling and Terry Farrell]. . . . postmodernism in philosophy typically centers on a critique of the modern ideas of reason and the rational subject. It is above all the 'project of the Enlightenment' that has to be deconstructed, the autonomous epistemological and moral subject that has to be decentered; the nostalgia for unity, totality, and foundations that has to be overcome; and the tyranny of representational thought and universal truth that has to be defeated. (Baynes *et al.*, 1987, 67–8)

The following is an example of these winds of change blowing in geography:

Modernism is the prevailing methodology in economic geography. It is the view that the only real knowledge is knowledge tested by rigorous scepticism. With knowledge defined as only that which cannot be doubted, it follows that metaphors with the connotation of ambiguity and imprecision are regarded by modernism as mere ornaments. At best metaphors are empty, conveying no additional information; at worst they obfuscate and detract from the 'facts'. . . .
In contrast, in postmodernism metaphor is viewed in quite different terms. Metaphors are an intrinsic part of language and knowledge; they are not simply decorative. This paper argues that by exploring metaphors used by economic geographers from a postmodernist perspective, one is provided with first, an understanding of both the internal and external relationships that shape economic geography, and, second, a critical foil to evaluate the appropriateness of the metaphors employed. (Barnes and Curry, 1988)

Relph (1987) has neatly rotated the evident contrast between modernist and post-modernist architecture into a model of spatial contrasts within and between urban landscapes. (Note: he uses the term 'Late Modernism' to describe modernist building constructed from about 1970 to the present in the declining phase of modernism.)

In a model case the landscapes of all these phases would be juxtaposed. There would, perhaps, be a modernist city core of skyscrapers and canyon streets, surrounded by post-modern districts of gentrified housing and warehouse/boutiques, then some 1920s garden suburbs with traditional retailing streets surrounded in turn by a band of corporate suburbs laid out in neighbourhood units, split by arterial roads and shopping malls, and then a ring of sleek late-modern office buildings and electronic plants lining expressways. . . . Most new towns are wholly in the institutionally planned, ordinary modernist style of the 1950s and 1960s; . . . they have in fact a one-dimensional landscape. (pp. 241–2)

Relph cites Carmel in California and the southern hill towns of France as examples of places which have escaped the 'high modernist phase' and so have passed straight from traditional forms to the somewhat similar profiles of post-modernism, symbolized by the banishment of flat silhouettes in favour of the gabled roof and the ornamented facade. Another example of this last case is to be found in Salisbury, England where, paradoxically, the rather modernist (because city-wide) planning fiat, preserving view of the cathedral spire from all compass points, effectively precluded the high-rise profiles of the high noon of Modernism.

A keyword in post-modernist approaches is 'deconstruction' (Dear, 1986, 372–3; 1988. 266–7) as a retreat from over-arching concepts such as those which could be labelled as some form of structuralism; but, almost in the act of deconstructing human geography, there are urgings for the subject's reconstruction (Dear, 1988). One can also see that rejection of terms like 'over-arching structures', 'totalizing discourses', 'metascience', 'meta-narrative' by post-modernists would lead to a confrontation with Marxism. J. Graham (1988), writing as a Marxist geographer, has dramatized this confrontation, as follows:

Using the oppositional rhetoric that characterizes the most obstreperous post-modernism, one might summarize the post-modern point of view on Marxist geography as follows: Marxism and class are dead, but geography and locality are alive and well. More generally, modernism, homogeniety, rationality, mass production, metanarrative, tract housing, and space are dead. Long live post-modernism, pluralism, power and desire, small batch production [i.e. post-Fordist flexible industrial integration], local narrative, indigenous architecture, and place. (p. 60)

But if post-modernism suggests the avoidance of arbitrariness in knowledge, this might lead to an emphasis on methodological comparisons, and these are offered in Chapter 8 below, and extended towards epistemology in Chapter 9.

Modernism and post-modernism are respectively exemplified by the continuing development of Habermas's theory of communicative action (explained in Habermas, 1971, 92) and Giddens's structuration theory. It is the second that has had greater influence within human geography, and Gregory (1992) has taken up the specific problem of time–space distanciation in structuration: 'Social practices reach beyond the here and now to include interactions with others who are absent in time and space. . . .

Giddens calls this process of "stretching" time–space distanciation.' Gregory argues that Giddens consistently theorizes spatiality in terms of power and domination. While this aspect is of immense importance—characteristically it is absent from most 'interpretalist' treatments of space—Giddens in practice allows little conceptual room for the efficacy of 'signification' and 'legitimation', or, more simply, 'culture': 'conceptions of space are not mirrors but *media* of time–space distanciation: [and] . . . time–space distanciation is structurally implicated in the time–space constitution of human subjects.' (The full argument is to be found in Gregory, 1992, ch. 7.)

In the case of transcendental realism's abstract research, structure is allowed to impinge on something specific in time and space, via the actions of knowledgeable decision-makers. But in transcendental realism's concrete research, where specific political situations and events are allowed to impinge inductively on structures, this would be a poison pill in reverse for Marxism, if the structures were held to be historically materialist in origin, with all the usual Marxist claims to holism. If I say that it seems that Marxism cannot take too much realism, I hope you will not misunderstand. It remains true that realist approaches are much more materialist than idealist and that the materialism often has strong Marxian overtones or undertones. In that case transcendental realism seems insufficient to paper over the ideological cracks. Perhaps thoroughgoing historical materialists will come out and say whether or not 'materealism' (*sic*) is revisionism.

Cochrane (1987, 354–5) has noted two forms of criticism of transcendental realism in a squeeze play from opposing political perspectives. First, it has not completely broken with structural (Althusserian or scientific) Marxism, but offers instead a more flexible structuralism 'in which it is agreed in principle that actors are important, but in practice it is difficult to discover what their importance is' (p. 354). The structuralist apparatus remains essentially in place, and we are advised to liberate ourselves from this 'orthodoxy' by paying more attention to the 'new right' (Saunders and Williams, 1986, 393–4, 399). Second, transcendental realism is seen as a retreat from Marxist theory:

Instead of insisting upon the rigour of historical materialism, Sayer proposes a realist philosophy. . . . The problem with this superficially attractive method is that there is nothing within it, apart from the judgement of individual researchers, as to what constitutes a special instance to which special processes inhere or as to what contingencies (out of a potentially infinite number) ought to be taken seriously. There is nothing, in short, to guard against the collapse of scientific understandings into a mass of contingencies exhibiting relations and processes special to each event. (Harvey, 1987, 373)

Yet another criticism comes from N. Smith (1987, 65–6): 'theory and empirical investigation are wedged apart and what began as a search for the middle ground ends by making the knife edge so sharp it is unwalkable'. In the course of a critique of Massey's (1984) *Spatial Divisions of Labour*, Cochrane (1987) seems to agree with this:

The retreat from grand theories of structural Marxism with their search for dominant archetypes is real enough. Unfortunately, it seems to have been replaced by an almost equally unhelpful search for structural relationships at the micro-level—by a search for necessary relations where none exist. . . . If each place is unique, then each conceptualisation seeking to explain industrial location ends up having to be unique, too. And, as a result, there is also an increased stress on the uniqueness of individual firms or capitals. Each type becomes effectively insulated from the other. (p. 361)

Structuralism has here run up against the problems of the unique that so bedevilled discussions about geography as a science before 1966 (see references to Bunge, 1966*b* in Chapter 1 via Index; and N. Smith, 1987, 66). But, unlike what happens in the hypothetico-deductive method, case-study lessons (leading to problems) cannot be used in structuralism as precursors to modified or enlarged structures (cf. theories in scientific method) in a constant-revision mode; this is because a-priori structures are brittle, fixed in place to provide theory for the empirical facts. If what we learn as we progress to further problems stretches the structures too far, they become unstable.

Both structuration and transcendental realism attempt to provide social science in general and human geography in particular with one complete, ready-made system with which to confront all empirical cases—like a recommended blueprint to which nothing much can be added or subtracted without the device to which it refers falling apart. Fig. 14 attempts to show that a theory which, in application, is always useful, to some extent does not distinguish between case-studies, nor does it provide strong guide-lines

Let A, B, and C be particular theories,and
X, a more general theory, applied to
Case-Studies I — IV and so on to n

Case-Studies

Theories	I	II	III	IV $\longrightarrow$ n
A	√ x	x	x	x x
B	x	x	x	x x
C	√	√	x	√ √
X	√ x x √	√ √ x x	x √ x √	x x √ √ √ x √ x

Let √ and x represent those parts of
case-studies 'explained' and 'not explained'
(respectively) by theory in question.

Resultant questions

Theory A : often disproved — What is special about Case-Study I? Can theory be improved?

Theory B : always disproved — Discard. What might be a replacement theory?

Theory C : occasionally disproved — What is special about Case-Study III? Can theory be improved?

Theory X : a general theory has something correct and something wrong in application to every case-study — What are the questions raised? Or, where do we go from here? (If theory amended to fit more of some cases, may no longer be a general theory.)

FIG. 14.   *Relative patterns of success and failure for four different 'theoretical theories', and questions raised as a consequence of their applications to empirical data within case-studies*

for its own improvement. Jackson and Smith (1984, 204–8) attempt to grasp such a theory by conflation: James's and Dewey's interactionism, Simmel's conflict theory, Weber's stratification, and Giddens's structuration. This is getting close to betting on all the horses in a race, but, while producing a winner every time, there may well be net loss, and the system does not teach us how to maximize profits and minimize errors in future specific races (or cases); or, indeed, how to imagine how the system might be improved for particular case-studies, nor to distinguish between case-studies that are appropriate for empirical application from others where difficulties outweigh advantages.

Human nature can be very perverse. We search and search for ever more 'perfect' and comprehensive solutions to our academic problems. And if we ever found such a system, we would condemn future generations of students to its mere application by rote. Geography students of the future need have no fears on this score.

Afterword to the parade

> 'Well, nobody's perfect.'
>
> (Very last words, spoken by Joe E. Brown, in Billy Wilder's
> film *Some Like it Hot*, 1959)

So the parade of -isms in geography has passed by, with no doubt more to come in future years. This demonstration of disadvantages in every available theoretical system has been provided not through malice aforethought but because disadvantages are a feature of every human endeavour; and proponents of various systems within geography have almost universally been very reticent about the shortcomings of their brain-children. We consumers are seeking that which gives us the maximum *net* advantage in solving our problems, and to do this we must rely on our own critical judgement when applying concepts and methods. Fortunately, every exercise of critical judgement strengthens our capacity to call on that faculty, even if our judgement eventually proves faulty, for we certainly learn most from our mistakes. The following chapters pass on to other concepts and methods that have engaged the attention of geographers, or which have impinged on geography from association with other disciplines: behavioural geography, systems, creativity, perception and the 'real world', and the question of changes in geography. A continuous exercise of wary judgement will still be necessary. Skip to the last chapter if you wish your present guide to disclose his own methodological and epistemological stance, and see if your critical judgement is strong enough to spot his mistakes. Meanwhile, we have to remember that we have not only to sort out our own ideas about the world but also to try to understand how significant actors (spatial decision-makers) have interpreted the world in which they act. This extra dimension accounts for the rise of behavioural geography.

6

Behavioural Geography

Reasons for the rise of behavioural geography

Goodbye for the moment to epistemological bases or particular philosophical perspectives because behavioural geography came into existence through a problem-led approach. Geographers have become progressively more interested in processes: how phenomena function in the case of physical geography; and, in the case of human geography, how people behave to produce the spatial patterns that can be mapped. Some relationships between behaviour and other concepts are shown in Fig. 15. A concern with human behaviour leads happily into links with sociology (group behaviour) and psychology (at the individual level)—rather less happily into that prickly thicket of the infinite regress. If 'reduction' is used to refer to ever more basic explanations, then it seems

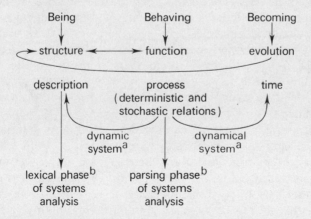

FIG. 15. *Behaviour and related concepts*

[a]The distinction is by Thornes (1987): while dynamic systems function, the 'dynamical' approach studies how a system behaves over time.

[b]After Huggett (1980, chs. 4 and 5).

TABLE 19.   Five stages of psychological reduction

5: spatial pattern, function; social pattern, structure
   4: behaviour
      3a: values and/or attitudes[a]
      3b: motivation to comply with norms[a]
      3c: normative beliefs[a]
         2: psychological drives
           1: brain processes

---

[a] For a discussion of these as 'three kinds of variables that function as basic determinants of behavior' and for readings in attitude theory see Fishbein (1967). It is useful to follow Meddin (1975) and regard an attitude as directed towards concrete objects in the environment as opposed to value, which has a more abstract connotation.

Source: Bird (1977a, 22).

respectable, and appropriate, to reduce down as far as level 3 in Table 19. If we go back to the positivistic geography of the 1960s, the level of explanation sought for behaviour in Table 19 was '3b: motivation to comply with norms', and the norms were economic optima. Here was the realm of economic man. This being was endowed with necessary knowledge and intelligence, to seek the optimum goal of maximum profits, a superman, easier to place in the context of macrostudies where deterministic regularity of behaviour is most comfortably predicated. Before humanistic geographers object, let it be pointed out that in many areas of human action regularities can be relied upon. Life assurance brokers can predict an individual's likely age of death based on actuarial tables, and have to do so with profit on the quoted premiums to stay in business. So two big issues reared up at the birth of an overt behavioural geography: optimization and determinism.

Behavioural geography cannot flourish if any deterministic forms of land–man relationships are believed in. And what has been called 'mild environmental determinism' did linger on for a long time (Bird, 1983a, 58–60). In the 1950s there was even a compulsory paper in the University of London degree entitled 'The physical basis of geography' (the title of a highly successful geomorphological textbook, Wooldridge and Morgan, 1937). And Martin (1951, 8) clearly stated his position in a paper entitled, 'The necessity for determinism': 'the older determinism was apt to over-simplify by skipping several links in the chain of cause and effect. From climate to civilization direct is too big a jump, but we would all recognize that *through its control of agriculture* climate does have a considerable indirect effect on civilization' (my italic). Seventy-two pages beyond Martin in the same journal occurs a paper with the title 'Weather–crop relationships' (Frisby, 1951). Up to the mid-1950s there was a widespread view in Britain that the land was the primary geographical document, and it seems so 'natural' to think of physical features affecting man's activities that many still slip into thinking that way today; yet there has existed a devastating counter-argument since 1948:

... milder propositions of environmentalism ... still linger on ... there is no proper natural way of shaping life but innumerable ways, not sorted out by nature but reduced by man's choices past and present. People live differently in similar environments and differently at different times in the same environment, without feeling any environmental pressure to lessen these differences ... Signs point to the most promising approach in every case as beginning with man's dynamic pattern of occupance in its total setting. (Platt, 1948, 351, 352, and 356)

Let me resurrect an almost forgotten paper by K.G.T. Clark, published posthumously a year before Martin's, and, ironically enough, prepared for publication by him. In the quotation below, Clark (1950) is using the example of rice cultivation:

No possible combination of physical conditions ... could in itself give rise to the cultivation of rice.

This can only occur when the mind of man conceives it worth while to grow rice, and when he sets into operation a series of events which culminate in the crop. Thus it is nonsense to consider the physical conditions as causal: they merely form a relatively stable medium of necessary conditions within which the true, human, causal factors operate. . . .

When men choose a course of action they are choosing from among what, relative to their own mental and material equipment, *seem* possible courses. (pp. 17 and 21)

Not only did the rise of the behavioural approach finally extinguish environmental determinism, it also broke the exclusively 'vertical' land–man relationships by adding in reasons for spatial patterns that may arise outside the area. Hodder (1965) adduces an external stimulus for the rise of periodic markets in Africa south of the Sahara (confirmed by R.T. Jackson, 1971, 31 in southern Ethiopia). Exogenous forces have often been neglected by geographers, as in too exclusive a reliance on the endogenous forces of central place theory. Fleure (1947) did not fall into this narrow world: 'we are dealing with a being who can transplant experience in a unique degree. Neither the bastide towns of Wales nor the German cities of Transylvania are in any real sense the children of their immediate environments; both result from deliberate transplantation of experience gained far away' (p. 30).

The positivists of the 1960s were great devotees of normative geography. Their revered forerunners were von Thünen (anticipated by Adam Smith, see Chisholm, 1979), Alfred Weber (1909), and, for settlement and market area development, Christaller and Lösch (see the basic thrust of Haggett, 1965). The von Thünen and Weberian perspectives are mirror images of each other as, respectively, the quest for the optimum activity for a given location and the optimum location for a given activity. Christaller's central place theory is the optimum geometric pattern for an isotropic plain. The chief difficulty arising from this isotropic plain is not the annihilation of physical geography but the fact that it is not connected to behaviour anywhere else in the world by land or sea. The striving to understand optimum patterns seems beautifully pure, until it is realized that several factors vitiate the enterprise, as Simon (1952, 1957) first pointed out in his substitution of satisficing bounded rationality for optimization. Empirical examples will always fall short when compared with optimum models which are normative (i.e. attempting to establish norms as a datum). As more satisficing components are added to the model, the diminished 'noise' created by the case-study becomes a more interesting residual, making possible sharper repercussions on the model used in the first place. Table 20 attempts to list some of the main reasons for behaviour which embodies compromises away from, or down from, optimum objectives; and the sources for this table are to be found in the whole developing field of behavioural geography over the last twenty years. Certainly, two of the earliest to apply these ideas were Kates (1962) and Wolpert (1964, 1965, 1970), who saw the necessity to go behind the overt act to get at the reasons for the actor's reasons. But, as Golledge (1981) points out, in allowing the emphasis to be placed on the actor, it does not follow that a return to the unique is being

TABLE 20. Fifteen reasons for satisficing behaviour

1. Imperfect knowledge.
2. 'Lag' of information on which to act.
3. Imperfect intellectual appraisal of problem.
4. View of choice within an unnecessary limited range of options (above are manifestations of 'bounded rationality').
5. May take short-run advantage over long-term greater gain.
6. Profit maximization only one of many goals in composite 'quality-of-life' aims.
7. Aspiration level may be low because of personality factors or
8. through low aspiration level of society.
9. Aspiration level of decision-maker depends on actions of competitors, current, and as estimated in the future = an emulation level (aspiration level for industrialists may be maintenance or increase of market share, or, put in a more modest and homely fashion, making sure that one's competitors are not doing better or stealing a march).
10. Decisions are based on the history of past successful decision-making, which creates a problem if there are unperceived changes in the 'task environment' (often a problem for older people).
11. When decision has to be made, all desirable options may not be available = choice is perforce from a restricted range.
12. Choice is affected by optimism or pessimism regarding the consequences of action and by the state of the economy as to whether or not short-run advantage is to be taken at expense of possible long-term gain (see 5 above).
13. Propensity to take risks in situation where risk-taking seen later to have effect on outcome (refers back to 7).
14. Communities are not made up of one-person decision-makers, but decisions often made in the context of households, not identical in needs, tastes, and incomes, within families or between families; and in organizations, decisions made in the context of the internal power situation as well as in response to problem faced.
15. External constraints internalized, e.g. perceived social norms of standard of opportunities and 'acceptable' behaviour (see Figs. 17 and 18).

advocated, because we can aggregate actors into groups with similar motives and aspirations, rather than aggregate people as data in arbitrary units such as census tracts. Not only is emphasis placed on the actor, but also on a being who is no mere observer of the passing scene, but who is, and always has been, a creator of it. After all, it is geographers, past and present, who have created what we call 'academic geography', which is the study of the world created by people:

> . . . whereas [others] think that the earth made man, man in fact made the earth. (Letter from Marsh, 21 May 1860, quoted by Lowenthal, 1965, p. ix)

Man is not *homo percipiens* but *homo creator*. . . . Although its image of man remains mainly passive, cognitive psychology does make a useful contribution to our investigation. It shows us that perception geography overstresses the reception of stimuli, the registration of facts, thus emphasizing the passivity of mind, and pays too little attention to the construct, the active cognitive behavior that creates information. In other words, perception geography emphasizes the influence of the environment on man and neglects the counter movement. (van der Laan and Piersma, 1982, 417)

It seems unavailing to ask for a definition of behavioural geography, because it is not a discrete subfield of human geography. A more profitable position is to regard it as a perspective which emphasizes the importance of human behaviour (see also Golledge and Stimson, 1987) in helping to explain the reasons behind spatial relationships. In stressing the actor, behavioural geography tries to make him more lifelike. The question arises as to how far back should we trace the reasons for the reasons. To require geographers to delve into brain processes is obviously excessive, but digging a little deeper than 'economic man', with the aid of psychology, has surely already enriched our contribution to the understanding of spatial patterns and organization.

Cognition, images, mental schemata, and the case against mental maps

Behavioural geography may not connote any particular epistemology, but it cannot flourish under all geographical perspectives. We have already seen how positivist geographers relied on economic man, who was supplanted by Mr Satisficer. As Pickles points out (1986, 19), humanist geographers would not accept that laws of human behaviour can be established, although they would approve of the reporting of behaviour as explained by the actors themselves. A middle ground is based on the belief that, while human behaviour has very complex sources, certain recurring constituents exist with a demonstrable relationship between them. The counterpart of this geographical middle ground in psychology is some form of learning theory. Downs (1970, Fig. 8, 85) has produced an interesting diagram of these constituents of spatial perception and subsequent behaviour over space. His constituents are listed below, together with my running comments. We enter Downs's system at the 'real world':

1. 'Real world': I place this in inverted commas because the expression begs a necessary discussion, which I should like to postpone to Chapter 8.
2. Perceptual receptors: I prefer 'Cognition of the real world' for three reasons:
   (a) 'perceptual' is often held to be visual, though it may not be so exclusively;
   (b) 'receptors' sounds wholly inductive, 'rays of information' coming into the eye, perhaps the most commonly held popular view of the way we see; whereas vision is also deductive, perhaps even wholly deductive as some psychologists believe (psychological school of mental constructivism);
   (c) perception is a subset of cognition, and cognition is an amalgam of perception and 3 and 4 below (see Table 21).
3. Value system: derived from an implicit or explicit epistemology or religion, and includes attitudes to concrete objects.
4. Image: discussion of this below.
5. Decision: based on searching information available, not on searching real world as Downs's diagram suggests.
6. Behaviour: actions that actually take place and which, as far as geography is concerned, have some form of spatial expression in movement and/or pattern.

Images are deeply embedded in cognition and are often the sources of behaviour. Lloyd (1982) has performed the useful function of acting as a scout within psychology to see what geographers can learn from a sister discipline. He provides us with a summary of three theories, and bear in mind that what follows is a summary of a summary. First, there is radical image theory, in which the actor or image-maker would say to the investigator, 'Texas looks big.' Perception has provided a visual image, or mental map. This last phrase has acted as a trigger of interest for geographers. Second, we have the conceptual (verbal) propositional theory. I have

TABLE 21.   Suggested hierarchies of cognitive terms

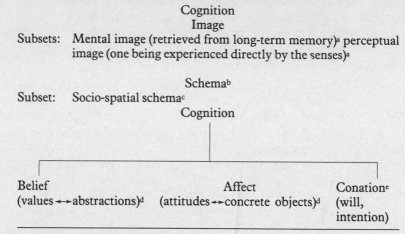

Cognition
Image

Subsets: Mental image (retrieved from long-term memory)[a] perceptual image (one being experienced directly by the senses)[a]

Schema[b]

Subset: Socio-spatial schema[c]

Cognition

Belief	Affect	Conation[e]
(values ←→ abstractions)[d]	(attitudes ←→ concrete objects)[d]	(will, intention)

[a] Lloyd (1982, 535–6, following Shepard, 1978).

[b] 'The schema may be defined as the cognitive structure or coding system that allows the individual to respond appropriately to a shifting pattern of environmental stimuli' Gold (1980, 41).

[c] Lee (1968), see accompanying text.

[d] Following Meddin (1975).

[e] The eye . . . sees only what it seeks' (Darby, 1962a, 5, quoting Marsh, 1864, 15).

added '(verbal)' to the title for the sake of exposition: here the image-making depends very largely on semantic propositions. Our actor would say, 'Texas is big.' Here abstract conceptions of Texas (proper name) and the idea of 'bigness' are compared and conjoined because of a perceived similarity. Experiments seem to indicate that children use the first, mainly visual type of imagery more spontaneously than the conceptual (verbal) propositional method, which develops as one matures. Conversely, such maturation leads to atrophy of the direct, visual 'picture-imagining', and this helps to explain the importance of children's imagery and naïve art. The third theory, the dual coding theory, combines the first two theories and asserts that, for pictures and concrete ideas that are easily visualized, we use a form of the radical image theory; but abstract material is conceptualized by the conceptual (verbal) propositional method. Fig. 16 attempts to show this dual coding theory in diagram form, leading to four

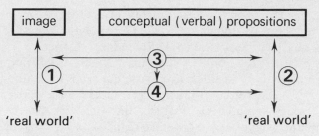

FIG. 16. *An interpretation of the dual coding image theory*

Note: This interpretation is by the present author. Double-headed arrows represent four comparisons between two entities or two earlier comparisons, as follows:

1 'Real world' compared with stock of images, or what can be visually imagined, as in the radical image theory;

2 'Real world' compared with stock of conceptual (verbal) propositions, as in the conceptual-propositional theory;

3 The comparisons are themselves compared to see which gives the faster or more 'sense-making' solution, compared with past experiences and/or mental schemata. (Bear in mind that such comparisons may take only milliseconds.) Adults' conceptual (verbal) propositional stores become richer as they mature, and this explains the increased relative use of this method as one gets older.

4 The results of these three comparisons are then used to produce a composite image in which visual images and conceptual (verbal) propositions are combined in varying proportions.

Source: The theory originates with Paivio (1969, 1971), summarized for geographers by Lloyd (1982).

'comparisons' which, of course, may take place in milliseconds and largely unconsciously.

The psychologist Lee (1968) has introduced the concept of the socio-spatial schema as a result of an investigation of residents' cognition of neighbourhoods in Cambridge:

People move about the local urban environment to satisfy a wide range of needs with minimum effort. The continual locational coding that arises from this activity precipitates in the form of a socio-spatial schema which, in turn, governs future navigation and movement. Each schema is unique, but is related in lawful ways to the physical environment and to the personality of its possessor.

Consentaneity of schemata occurs in varying degrees and its measurement provides a means of predicting behaviour for a given aggregate of people with a territorial base. (p. 263)

Another strand in cognitive theory of interest to geographers was generated by Lynch's (1960) *The image of the city*, where perception is structured into 'paths, edges, districts, nodes, and landmarks'. P. R. Gould (1966) and Gould and White (1974) attempted to make manifest group mental maps by asking people to rank places according to their preferences. For a while, mental map studies sprouted everywhere, but the approach began to wither during the 1970s. 'Mental maps' is still a very popular seminar topic for studen:s to write up. Initially, they are puzzled by the decline of such studies. Part of the reason for this decline is that the approach is somewhat idealist and began to wane as materialism gained ground in human geography. We can see this in action in Jensen-Butler's (1981) criticisms of cognitive mapping, and it is worth looking at some of his extensive comments in detail, because they excite comment in their turn.

Cognitive mapping was held to be 'pre-theoretic', in that the retrieved maps would form the data on which behavioural theory would be based. Jensen-Butler (p. 25) rightly objects that no research can be atheoretic. Remember 'theory-laden observation'. The mental maps are recognized as being different from the real world. But how does the investigator establish what is the real world datum (Jensen-Butler, p. 29; see also Bunting and Guelke, 1979, 454)? One answer might be that it is legitimate to compare respondents' maps with surveyed maps. But maps based on Euclidean coordinates are often misleading because they ignore time-distance, cost-distance, and time–space convergence, a most important concept introduced to geography by Janelle (1968, 1969) which demonstrates that improvements in transport have proportionally greater impact the longer the distance of the transit. For example, he shows

that an increase in transport-speed on a route connecting successive places will generally result in greater convergence rates for more distant places than for the closer or intervening ones . . . the relationship of time–space convergence with distance for any given transport improvement is a factor which helps to augment the dominance of the higher-ordered centers . . . (Janelle, 1968, 8–9)

. . . human behaviour is affected only by the portion of the environment that is actually perceived . . . (Gould and White, 1974, 48)

'. . . patently wrong', says Jensen-Butler (p. 29), and it is hard not to agree. Endogenous variables within the subject, such as age, aspirations, goals, norms of behaviour, and perceived constraints are allied to and affect perceptions of the exogenous environment. Often cognitive mappers treat the respondents as rather passive receivers of information. The questions put to them in the groundwork for deriving mental maps are often posed in 'a fantasy world where they are suddenly free to choose' (Jensen-Butler, p. 35). Mental maps are often interpreted in relation to some of the attributes of space, but these are 'hopelessly confounded with other properties of objects' (p. 36). A high reliance on variables which vary most closely with the map leads to 'rationalization *post factum*, and there is no guarantee that the explanatory variables identified by the investigator are important for the respondents (p. 37). This is all well put, and it is an attack that cognitive mappers have now to take into account. But this attack on cognitive mapping is part of a wider materialist attack on idealist work in geography, as will be shown by two final quotes from Jensen-Butler (on two statements by Downs and Stea):

Organization is best expressed by the idea of making sense out of things, by the effort after or the search for meaning. Sense (organization) is not given by the environment out there. (Downs and Stea, 1977, 83)

The idea that order is a property of mind and not of the real world is a basic idealist postulate, negating the existence of a real world, governed by developmental laws, existing independently of the human mind. This implies that the concept of objective knowledge of the real world must be given up . . . (Jensen-Butler, 31)

The materialist view of the development of consciousness places man's interaction with his social and natural environment at the very centre of the analysis . . . (Jensen-Butler, p. 34)

Jensen-Butler goes on immediately to state the epistemology of historical materialism. But beware! I have paraphrased some thirteen pages of his argument, and used selective quotations. The aim has been to illustrate another facet of the idealist–materialist debate, so important in human geography. At least conflicting epistemologies have one advantage: they act as a spur for the uncovering of faults in methodologies based on philosophical foundations with which the critics do not agree.

Behavioural geography: a possible agenda

So far in this chapter the *raison d'être* of behavioural geography and links with psychology have been outlined. This helps to define a perspective which can be turned on various subfields of human geography and also on applied physical geography. This is the basis of the two-part agenda of topics for a short course given in Table 22. The included items cannot possibly exhaust what the perspective has to offer, so a starter list of possible additional topics for seminars has been added as a third constituent of the course. Goodey and Gold (1985, 588) criticize such 'stand-alone' behavioural geography courses; yet perhaps they would approve of the attempt to engage the student's 'sense of geographical exploration as part of the teaching programme' (ibid.). Whereas behavioural geography can act as an enrichment and critique of many fields of geography, in turn it has not escaped criticism. Be assured that your guide will not neglect this aspect before leaving the subject, but meanwhile space permits only two other topics from a wide agenda: the problem of behavioural constraints and a suggested method of probing the decisions of busy decision-makers.

TABLE 22.   Behavioural geography—possible agenda for a short course[a]

Part 1: Theoretical background
  1. The rise of behavioural geography
  2. Behavioural sequence—cognition
  3. Behavioural sequence—perception
  4. Spatial cognition
  5. Milieux, phenomenal environment, behavioural environment, operational environment

Part 2: Behavioural studies in geography—a wider perspective
  6. Hazard perception
  7. Behaviour in agricultural geography
  8. Behaviour in urban areas
  9. Behaviour in industrial geography
 10. Behaviour in transport geography
 11. Time geography
 12. Landscape assessment
 13. Imposed landscape behaviour
 14. Critiques of behavioural geography

Part 3: Some possible seminar topics
  1. Landscape perceptions in art
  2. Landscape perceptions in literature
  3. Critique of 'mild' environmental determinism
  4. Imperfect knowledge of geographical actors
  5. Mental maps
  6. Perceptions of centres
  7. Perceptions of peripheries
  8. Environmental stress
  9. Idea of 'The Frontier'
  10. 'Culture' in geographical studies
  11. Tourists' perception
  12. Thematic leisure landscapes
  13. The Americanness of the American cityscape
  14. A geography of voting
  15. Evolution in the work of a noted geographer[b]
  16. The generational model in geography[c]

---

[a] For a possible longer course see Golledge (1985, 122–5), and Golledge and Stimson (1986).

[b] Possible names are: Bunge, Gould, Harvey, Olsson, because their work published so far has undergone marked changes of perspective = 'the changing behaviour of "geographical actors" '. Such changes are not necessarily to be viewed as deserving either praise or blame. Bandwagon promotion can be a very valuable contribution; bandwagon hitch-hiking is not so estimable, as, by the time they can be recognized as truly rolling, bandwagons are generally rolling downhill.

[c] The idea is to exemplify and comment on Johnston's generational model (Johnston, 1978, 1979*b*) as describing, or not, the behaviour of professional geographers: 'What is being suggested . . . is a generational model of disciplinary development . . . The external environment creates the conditions which are favourable for a redirection of scholarly effort. An iconoclast (sometimes several), *usually with secure position in the academic career system*, reacts to these conditions by creating new exemplars, most often within the established disciplinary matrix, and attempts to obtain the resources which will allow initiation of research programmes based on his ideas. If successful, his followers may form a branch, or constellation of branches, which either comes to dominate the disciplinary matrix, or, by revolution, replaces it' (1979, 184–5; my italic).

Behavioural constraints

The constraint of bounded rationality (Table 20) helped in the rise of behavioural geography, which has always been concerned with the limits on action. Materialists certainly stress the importance of real-world structures, whereas more idealist-oriented geographers tend to 'see' the structures through the cognition of the actors being studied. This is especially the case in European research—individual behaviour subject to group constraints,˜ societal and institutional, compared with a North American˜emphasis on *active* individual decision-makers, a contrast made in Thrift's (1981) survey. We may note here the adoptive/adaptive dichotomy, where 'adaptive' signifies active decision-making in contrast to being passively 'adopted' by the environment (as in the successful products of natural seed distribution; see Alchian, 1950; Tiebout, 1957; Krumme, 1969). We have seen that historical materialism presupposes constraints over individuals, and this was also the emphasis in the earlier versions of Hägerstrand's time geography—a convenient summary of which is to be found in Pred (1977). The aim of the model is not to explain behaviour but the constraints on behaviour:

1. Capability constraints: time must be allocated to physiological activities (sleep, eating, personal care) and time to move via available transport;
2. Coupling constraints: determining where and for how long an individual must be with others to produce, consume, or exchange;
3. Authority constraints: laws, economic barriers, power relationships limiting access to specific areas at specific times for specific functions (including the indivisibility of individuals, who cannot be in two places at once [telecommunications can overcome this] or do two things at once) (Hägerstrand, 1970; Pred, 1977, 209).

Once again we can summarize Jensen-Butler's (1981, 43–55) criticisms:

1. Hägerstrand has argued that his approach is based on physical realism (1970 paper in Swedish, quoted in Jensen-Butler, 1981, 45), but physicalism 'must be rejected because it is idealist, it assumes an identity between subject and object. . . . only observable elements enter into the structures considered, which . . . renders the concept non-materialist' (Jensen-Butler, 1981, pp. 46 and 47).
2. The model is only concerned with static structures (ibid. 47).
3. 'Space and time constrain the individual because they are

properties of objects and relations in the real world, not via their innate properties' (ibid. 47).

4. Pred says the constraints in time geography 'Specify the necessary (but not the sufficient) conditions for virtually all forms of interaction—social and otherwise—involving human beings' (Pred, 1977, 209). Upon which Jensen-Butler (1981) tartly observes: 'What the model in fact specifies are sufficient conditions for interaction to take place, assuming that it would if the conditions were met' (p. 49).

5. Changes in the constraints (innovations perhaps due to technical changes) come from outside the model so that the 'key explanatory variable is abstracted out of the model' (ibid. 50).

6. Only consumption is considered: the sphere of production is ignored.

7. Passive observation of behaviour is the method, without causal hypotheses (ibid. 51).

8. The final comment is on Hägerstrand's later methodological approach based on the compilation of individual biographies: 'The biographical technique is used idealistically because the correct specification of the relationship betweeen subject and object is not made in the approach. The "real" activity is compared, by passive observation, to the subjective elements of the biography. If they correspond, the "subjective" is assumed to "explain" the "real", as was the case in cognitive mapping.'

So the materialist has struck again.

However, comments 5 and 6 above have little to do with a materialism–idealism opposition, and have been developed by Brown (1981) in relation to innovation diffusion research: 'From the time geography perspective, . . . innovation diffusion research of the past has tended to be narrow in its focus upon single innovations and communications related to adoption' (p. 20). Brown has shown that the adoption perspective, which is the demand aspect of innovation diffusion, must be supplemented by three other perspectives:

1. The market and infrastructure perspective: the focus is upon the process by which innovations and the conditions

for adoption are made available to individuals or households, that is the *supply* aspect of diffusion (p. 7);

2. The development perspective: this examines the impacts of the innovations, not all of which are beneficial in all their consequences, and also the development stage of the environment in which the innovation diffusion takes place;

3. The economic history perspective: this embraces the invention and innovation processes and the preconditions for their diffusion, thereby avoiding the comment that the key explanatory variable is left out of consideration.

Constraints on spatial behaviour are now seen as highly complex, far removed from the concept called 'consumer sovereignty' in Walmsley and Lewis's review of work up to 1980 (1984, 87–8). Consumers are subject to many sorts of constraint: such as time, restricted trading hours, transport availability, lack of mobility (low income), age (possible low income and possible lack of car), quality and distance of shops (supply factors). But there are many other factors besides age that are rooted in the individual. Desbarats (1983) has studied constraints on behaviour in great detail. What she discusses is how actual behaviour decisions result from constraints on an individual's perception of what he believes to be optimal behaviour, and the constraints are both exogenous and endogenous to the individual (Figs. 17 and 18). In the context of spatial choice and constraints, we have the two familiar positions:

Those who view spatial behavior as a reflection of human free will expressed through rational choices tend to seek explanation at the micro level and to engage in disaggregate research. Those who stress the determination of human behavior by the objective forces of social dynamics tend to seek explanations at the macro level and to shun disaggregate research . . . Modes of explanation emphasizing system variables, however, are not necessarily incompatible with modes of explanation emphasizing individual variables. The two types of variables are intricately linked, as it is the differential susceptibility of individuals to constraints that mitigates the effect of system variables. (Desbarats, 1983, 353)

Perhaps we ought to pay attention to the types of spatial behaviour we wish to understand: repetitive 'enforced' behaviour (such as journey to work) might yield better to aggregate methods than

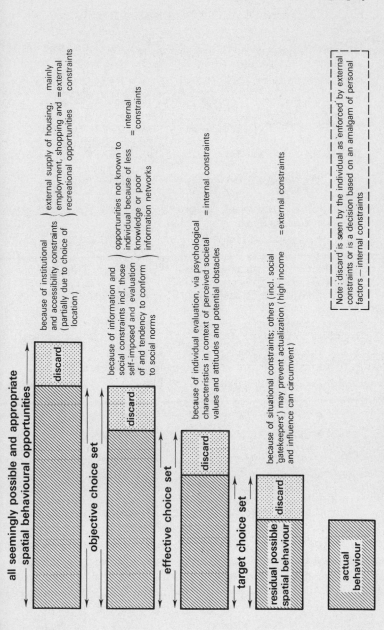

because of institutional and accessibility constraints (partially due to choice of location)

external supply of housing, employment, shopping and recreational opportunities — mainly =external constraints

because of information and social constraints incl. those self-imposed and evaluation of and tendency to conform to social norms

opportunities not known to individual because of less knowledge or poor information networks — = internal constraints

because of individual evaluation, via psychological characteristics in context of perceived societal values and attitudes and potential obstacles — = internal constraints

because of situational constraints; others (incl. social 'gatekeepers') may prevent actualization (high income and influence can circumvent) — = external constraints

all seemingly possible and appropriate spatial behavioural opportunities

discard

objective choice set

discard

effective choice set

discard

target choice set

discard

residual possible spatial behaviour

actual behaviour

Note: 'discard' is seen by the individual as enforced by external constraints or is a decision based on an amalgam of personal factors — internal constraints

FIG. 17.   *Constraints that reduce all seemingly possible and appropriate spatial behaviour to actual spatial behaviour*

Source: Based on and adapted from text and Fig. 2 in Desbarats (1983, 350–2; Fig. 2, 351)

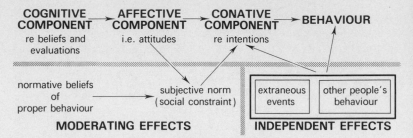

voluntary **ENDOGENOUS EFFECTS**

FIG. 18.   *A simplified version of Desbarats's 'conceptual framework for constrained movement behavior'*

Source: Adapted from Desbarats (1983, Fig. 3, 352): '. . . incorporates the internalization of the possible effects of extraneous events and the possibility of preventive action that such internalization implies'. The framework is derived from Dulany's theory of propositional control, modified by Fishbein in Fishbein and Ajzen (1975, 298–334; see Desbarats, 1983, 348, and Fig. 1, 349).

more discretionary 'one-off' behaviour (residential choice, annual vacation travel). But Pipkin (1981, 171) is hoping for unified accounts of the 'behaviour of consumers and producers, and of the mutual relationship between individual behaviour and aggregate spatial structures'.

Interviewing decision-makers: a little practical experience from behavioural research

In a Western-style capitalist mixed economy, certain individuals assume important roles as spatial decision-makers. Some have been described as 'social gatekeepers' or urban managers; others are the executive general managers of public utilities and private enterprises, large and small, who in the context of the power structures of organizations make decisions that have spatial effects such as infrastructure development and traffic flows and routeing. A humanist behavioural geographer would be in favour of letting these actors speak, but there are several practical difficulties. Whatever research methodology is adopted, there is a filter between us and the 'speaking data' through the process of theory-laden observation—every questionnaire structures the responses by the layout and nature of the questions asked. Even a passive operator with the stimulus card technique described below cannot avoid theory-laden apriorisms; but in this technique the respondents can escape via the open-ended format, though we are not guaranteed that the investigator will report all these escapes, some of which he may regard as 'irrelevant'.

In two research projects the cast of characters interviewed is given in Table 23. One advantage in asking similar questions of people in different organizations, or of people in similar

TABLE 23.   Spatial decision-makers in two European seaport research projects[a]

Aim: to study current and future problems of European Communities (EC) seaport development

Decision-makers (port directors and managers of enterprises)		Interviews	
		BID[b]	EUM[b]
In ports		20	20
In 'seaward customers of ports' (i.e. ship-owning organizations) as to their views of port operations, distinguished by decision-makers' location		20	20
In 'landward customers of ports' (represented by freight forwarders) as to their views of port operations and port routeing, distinguished by decision-makers' location	Ports	26	21
	Inland	10	15

BID = Britain, Ireland, Denmark; EUM = rest of EC (European mainland).

[a] Substantive results from the research projects are reported in Bird (1982, 1988) and Bird and Bland (1988); and there is more extended methodological discussion in Bird, Lochhead, and Willingale (1983).

[b] A hypothesis is implicit in this locational classification. In one case (BID), ports seek to follow what might loosely be termed a commercial concept, endeavouring to give efficient service at least cost to users, consistent with their meeting some defined financial requirements; and this meeting of financial requirements is normally, though not necessarily, defined in terms of the relationship between the income and the outgoings of the port undertaking. A totally different concept (prevalent in EUM) relates to maximization of the benefits and in so doing also takes into account benefits generated by, but outside, the port sector, for example the generation of employment and incomes. The hypothesis was that there would be different answers from respondents in the two areas of the European Communities.

organizations in a significantly different class of location, is the opportunity set up for comparative analysis of results. The two projects were designed to move on from investigator-focused studies of European seaports (e.g. Bird, 1967; Bird and Pollock, 1978) to situations wherein people who help make decisions can speak through the admitted filter of the investigatory technique. This consisted of a questionnaire and a tape-recorded interview. The major difficulty encountered was in setting up interviews. Experience suggests that the smaller the organization, the more reluctant will be the top decision-maker to co-operate. He is so busy with the day-to-day running of the concern and making his living, whereas the chief executive of a large organization has powers of delegation to allow himself more 'thinking' time about longer-term issues. Busy executives will not bother with complicated forms and they need to have their minds focused quickly on the main problems in question. The simple questionnaire was really an insurance in case the interviews yielded poor results, and was very constraining on respondents though easy to analyse statistically. Results were also compared with the 'working hypothesis' which was produced by the investigator, who completed the questionnaire before any of the respondents, based on his a-priori expectation of what they would answer, and this in turn was based on stated reasons. The investigator is now in a 'no-lose' position: if the results of the probe bear out the expectations, that is reassuring, provided that the reasoning *en route* is the same; if the results are different, this may be even more valuable in providing new insights and questions.

The 'free' interview was, nevertheless, structured in a simple way, because without any such structuring comparative analysis would be very difficult. A tape-machine was used to record the responses to a succession of statements on six cards. The interviewer remained passive throughout, merely explaining the procedure and confining contributions to encouragement when needed, which was rare. The six statements were designed to focus attention on the major problems facing European seaports and their customers. They had been tested in pilot interviews, and were generally in the form of propositions upon which the interviewee was invited to comment; the simple 'establishing' statement of the first card was succeeded by progressively more difficult topics to the more overtly open-ended last two cards, where respondents were asked to speculate about the future. Undoubtedly, the simplicity of the technique contributed to the successful responses in almost every case. (This interview technique may be compared with the more in-depth tape-recorded interviews by Stafford (1974), but his sample was only six firms and their eight location decisions (p. 170). It is apparent that almost all interviewees when talking about a subject which is close to professional expertise are interesting, find the experience enjoyable ('Somebody out there is listening at last!'), and become more eager as the interview progresses.

For many of the respondents English was not their first language, and thirteen of the 152 interviews used had to be translated into English. The interview transcripts were constantly compared with the words of the 'proposition set', that is, the words on each of the six printed cards to which the interviewee was responding. Only those passages of the interview transcript that related directly to the propositions were included in the analysis. Any particularly striking phrase or idea was quoted verbatim. Each separate relevant comment on the propositions was coded under a summary heading. By having a single person code throughout, inter-researcher variations were avoided. It should also result in a high degree of consistency of coding and categorization throughout the analysis.

The selection of themes proved a more appropriate method than the dissecting of texts into individual phrases and even words. It was appropriate, first, because of the high proportion of non-British or -Irish interviewees, who might often use peculiar turns of phrase, and second, because of the purpose and aims of the project, namely the identification of attitudes and perceptions of a set of fairly general issues, as directed by a few open-ended propositions. Accordingly, headings were raised to cover in summary form the relevant points appearing in the transcripts. Some of these headings emerge as 'major' responses. How can these be defined? Two similar points raised by respondents might be coincidence; three points that can be coded together under one heading certainly confirms some consensus, but others may consider the heading of minor importance. It was felt that, with eighty interviews in question in the first project, six or more points by separate interviewees spontaneously responding which could be coded under the same heading indicated something quite significant and gave the definition of a major heading. In the second project of seventy-two interviews, the threshold for 'major' responses was set at five.

There is, nevertheless, a methodological problem in this technique. The data supplied by the questionnaires and interview transcripts are on an individual basis and subjective—in a word, qualitative. Can they be analysed by anything resembling the hypothetico-deductive method? Some forms of content analysis attempt to safeguard as much objectivity as possible, thereby facilitating experimental replication by independent researchers. In the study reported here the size of the population was such that it was possible for one person to analyse all the interviews. An associated questionnaire did allow the responses to be analysed by simple statistical techniques but was very constraining on respondents. It appears that the form of content analysis used in research projects should be appropriate to the data in question, which may vary in the following ways:

1. Material gathered 'bespoke' for the research purpose or 'off the peg' from some other source (e.g. newspapers);
2. Circumstances of the interview, formal or informal, attitude of interviewer, method of recording information;
3. Type of probe: structured, improvisatory structured, or structured changing to 'free';
4. Topics investigated—narrow or wide focus;
5. Topics investigated—close to the interviewee (problems of subjectivity increase, but are present in all interviews) or distant; personal or impersonal (problems of confidentiality are not necessarily less when the topic is impersonal or apparently at a distance from the interviewee);
6. Topics investigated—recall from the past (may involve failures of memory and/or *post hoc* rationalization) or presentation of hypothetical problems (may evoke a 'hypothetical' rather than a real-life response if not constrained by actual real-life context). (For a summary of content analysis techniques, see Bird, Lochhead, and Willingale, 1983, 145–7.)

Content analysis appears always to involve two major steps: the breaking down of the interview to enable assignment to categories; and a 'vertical' comparison of the weighting or emphasis assigned to each category in each interview and/or 'horizontal' comparisons of similar categories and their weightings between interviews. We thus appear to have the relationship to the hypothetico-deductive method (a term borrowed from Medawar, 1969; see Chapter 1, above) outlined in Table 24.

We have to be aware of what we are doing. On the one hand, we wish to allow decision-makers to speak as freely as possible about the subject on which they are expert. On the other hand, we are obliged to focus their thoughts by the form of the questions asked so that their replies can be compared. We not only make comparisons between the responses but also with the investigator's apriorisms embedded in the working hypothesis. In these ways investigator bias during the field-work is acknowledged, and the fundamental preconceptions with which the work started are also acknowledged. When these are later modified, the evidence for so doing is provided by the 'speaking role' of the decision-makers as they help us elucidate the problems in question.

TABLE 24.   Hypothetico-deductive method and interview analysis

Hypothetico-deductive method	Interview analysis and conclusions
Problem	Problem
Working hypothesis	Working hypothesis, inc. questions and forms of probe in the interview technique
Data	Responses
Comparisons of data with hypothesis	Variant of content analysis, involving some form of comparison of categories within one interview, or between interviews, and any apriorisms within the working hypothesis

Criticisms of behavioural geography

From the moment that behavioural geography became recognized as an 'adjectival' subfield of the discipline it has attracted criticism. There appear to be three main reasons for this: divisions within geography itself, the problem of the infinite regress, and the tensions between micro- and macroscale studies. At least there seems to be a consensus that behavioural geography is not a subfield like biogeography or urban geography, where the subject matter is circumscribed by type of phenomenon; it is instead a perspective to be turned on the whole area of human geography and applied physical geography. As such, it can be practised from the positivist, humanist, structuralist, or radical standpoints. Each of these has criticized the others.

First came the criticism that behavioural geographers were merely propping up the positivist movement of the 1960s by providing the satisficing modification to optimization. As early as 1973 Rieser produced a scathing radical attack on behavioural geography's support for the status quo (p. 53), while Massey (1975) had a structuralist view of constraints: 'It is not simply that different historical epochs, and so on, will add or remove constraints on choice, but, to use a linear programming form, that the objective function itself will change. Goals and attitudes are just as much "a product of the system" as are the behaviour patterns which form their, however "imperfectly", derived actions' (p. 202). Later Cox (1981) complained that positivism in behavioural geography separated subject and object (p. 256), and when it became more humanistic there was lacking a theory of social life—an 'abiding weakness' (p. 266)—a gap which historical materialism is no doubt ready to fill. In a retrospective review Gold and Goodey (1984) noted the 'schism' within behavioural geography with mutual recriminations between positivists and humanists; with the former condemning 'deep subjectivity'; with 'casual ransacking of the arts and the humanities in search of often trivial insights'(p. 546; D. Gregory, 1981c, 2); and the latter rejecting the subject–object split and the fact–value dichotomy, reaffirming place instead of space (Gold and Goodey, 1984, 548).

To study behaviour is also to study the reasons for action, but Bunting and Guelke (1979) argued differently:

. . . geographers, we would argue, are not inherently concerned with images and other aspects of cognitive behavior but only with these as they represent a means to an end; i.e. a better way of understanding man's activity on the face of the earth. (p. 455)

'Unless you can use your image to do something, you are like a man who collects maps but never makes a trip.' (p. 456, quoting Miller, Gallanter, and Pribham, 1960, 2)

Our basic contention is that overt behavior is the primary concern of human geographers. A thorough description of human geographical activity would, therefore, seem to be the logical starting point of geographical analysis . . . It seems clear to us that the study of images and perceptions prior to, or at the expense of, detailed analyses of overt behavior is simply like putting the cart before the horse. (p. 458)

Three behavioural geographers replied in a debate immediately following the Bunting–Guelke paper and another 'reply' is to be found in Golledge (1981, 1327–8). But, in defending attempts to explain behaviour, one has to decide how far back the explanation must go—the problem of the infinite regress (see again Table 19). Svart (1974) had previously commented on the political dimension of what was later to be advocated by Bunting and Guelke: 'In so far as studies grounded in actual behaviour within given systemic circumstances preclude investigation of potential behaviour under very different hypothetical constraining conditions, they provide ideologically-biased information about the desirability of alternative social policies' (p. 303; 'absolutely correct', says Massey, 1975, 201).

Thrift (1981, 359) goes so far as to argue that behavioural geography 'is heavily circumscribed by constraints on what it can know given its emphasis on the individual'. Yet aggregation into groups can take place, and, in the case of repetitive travel, simulation of group behaviour becomes possible through modelling. But Couclelis (1986, 95) believes that such fundamental questions as 'how do people really behave?' and 'how do people really make decisions?' are not likely to be solved in our lifetime. Behavioural geography has certainly helped to demonstrate that the springs of spatial action are complex, and that constraints are both external and internal (see the Desbarats model in Fig. 18). It has also revealed something that we might have guessed at the start—that 'why' questions, which tickle our curiosity so fundamentally, are generally the most difficult to answer.

Yet this is not to say that 'how' questions do not lead into methodological complication as we shall now discover when we turn to look at the use of systems in geography and find that the emphasis of their application swings towards physical geography.

7

Systems and Geography: Their Diffusion and Doubts along the Way

Several strands

'Systems' no longer strike a talismanic chord in geography, but we shall have to judge whether this is because they have become so pervasive in the discipline or because experience of their use has run into too many difficulties. Undoubtedly, the various ideas nestling under the 'systems approach' first penetrated the subject via physical geography, and an early forerunner of the real push in the 1970s was the seminal idea of the ecosystem by the botanist Tansley (1935, 299). In a review of systems diffusion into physical geography, K. J. Gregory (1985, 140–60) has shown how soil scientists suggested an integrated systems approach which was later taken up by physical geographers. A similar importation occurred into climatology, and it is easy to realize how the atmosphere came to be regarded as an energy-exchange system.

A second strand involved two moves for a new metalanguage for science. Cybernetics was named in 1947 by Wiener, a mathematician, and Rosenbleuth, a physician. The word is derived from the Greek, *kybernētēs*, a steersman, probably to avoid the autocratic overtones of 'control'—a word we shall meet again in this chapter. Beer (1977, 151) has suggested that a modern definition might be: 'the science of effective organisation', which avoids description of any system in question, but refers to pragmatic qualities. A second attempt at a general scientific methodology was von Bertalanffy's (1968) promulgation of a general systems theory (hereafter GST). In this initiative, the thrust was to promote the general qualities of systems much in the manner of the general features of mathematics. Physical objects with their linkages and associated processes could be rendered into elements and linkages between them within systems. The general lessons of systems thinking are then applied to a particular configuration, and the output should provide insights into the original phenomena and their relational development (see also Mesarovic, 1964).

Let us assume for a moment that the systems methodology is firmly in place and that several component concepts have been understood and accepted. Systems analysis now supervenes. This operation was manifest first in control engineering and management studies, leading to various kinds of multivariate modelling techniques, developed appropriately in both physical and human geography. Notice that word 'control' again, which is sure to give us some trouble as we delve deeper. At this point we cannot avoid hearing the oft-repeated cry of the human geographer and other social scientists that the entry of man into a system throws its neat, mathematical-like logic off balance. We may here repeat Weaver's classification, now in the context of systems, which recognizes that, when minds are embedded in the system, there is a new set of problems:

1. Systems of *simplicity* containing a very few components which are directly related to each other;
2. Systems of *disorganized complexity* with large numbers of components with random or weak connections;
3. Systems of *organized complexity* with large numbers of components which are interrelated.

The classification is based on Weaver (1958, republished 1967), and there is discussion of the classification in relation to systems by Wilson (1981, 39–40; see also Thornes and Ferguson, 1981).

Pulling the strands together at this point, we note: that systems had an easier penetration into geography via physical geography; and that systems were proffered via GST as a ready-made tool-kit for problem-solving in all disciplines, including human geography. Here there was some resistance, partly due to the fact that problems there are mostly of the Weaver-type 3. The tool-kit analogy suggests that in systems we have no epistemological basis. They are not an expression of a particular philosophy but essentially a methodological aid. As such we can submit them to the test of pragmatism. Now there is a celebrated formulation by the founder of the American school of pragmatic philosophy, Charles Sanders Peirce: 'It appears, then, that the rule for attaining the third grade of clearness of apprehension is as follows: consider what effects, that might conceivably have practical bearings, we conceive the object of our conception to have. Then, our conception of these effects is the whole conception of the object' (1931–5, v, 258, first published in *Popular Science Monthly*, 1878). Here the 'object of our conception' is a system, and, to pass the test of pragmatism, it is defined solely according to effects 'that might conceivably have practical bearings'. This is a warning to geographers not to produce systems for their own sake, but only as a means to problem-solving. Chapman (1977), an advocate of systems in geography, frankly admits that he has not completely avoided this trap:

I have never been persuaded of what G.S.T. is trying to do, but I now find I have accidentally persuaded myself because it is what I have been doing. Hence, the book itself has become more and more ways of looking at systems, no matter their subject origin, and ways of thinking about systems, than a discussion of any particular system. (p. 404)

Now this is not to say that systems-building must be kept simple or not developed; systems may well become very complicated indeed as computers are able to handle more and more complex data sets. But for a geographer, *qua* geographer, systems-building is a diversionary exercise in another subject area—no doubt a necessary preliminary exercise, unless a ready-made and appropriate model is available. He may find it useful to conceive a system in order to formulate his research strategy, though I would not go so far as Chapman (1977, 6), who believes that as a 'framework for analysis it has no current peers'. The geographer may have to construct his own system because only he can recognize the interconnections betweeen the phenonmena in question. But he must always remember that the system is nothing but its own problem-illuminating output.

Some terms from systems

Many studies which use systems thinking extensively contain a glossary. Basic sources are Hall and Fagen (1956), Young (1964), and the geographers Chorley and Kennedy (1971, 346–59), who provide under 'system' no less than thirty different types. Langton's (1972, 128) preferred definition is by J.G. Miller (1965, 200): 'A *system* is a set of units with relationships among them. The word "set" implies that the units have common properties. The state of each unit is constrained by, conditioned by, or dependent on the state of the other units.' Langton equates 'constrained' with causal, 'conditioned' with functional, and 'dependent' with normative relationships. Some thirty terms are mentioned below in italics with hints of their implication in geography.

A basic difference is between *closed* and *open* systems. Every system studied by geographers is open in the sense of importing mass and energy, most notably from a proximate *task environment*, though geographical systems may expand into an appropriate *zone of opportunity*, and have transoceanic links via the atmosphere and intercontinental transport. Perhaps the greatest limiting assumption of central place theory is not that it is developed on a featureless plain but in a closed system without *inputs* and *outputs* via gateways (Bird, 1980, 1983b, for references on this theme) to the rest of the world. A *subsystem* is a functional component of a larger system which fulfils the conditions of a *system* (a structured set of objects or attributes with relationships between them), but plays a specialized role in a larger system. One lesson learned from systems analysis is that to improve a subsystem to optimum performance may be to the detriment of the larger system: flood-control measures in one area may exacerbate flooding downstream; savings through absence of fume-control measures in one city factory may pollute the whole urban environment. In *positive feedback* the work done by the feedback mechanism reinforces the main driving force. Sometimes this is known as *deviation–amplification*, producing *homeorhesis*—the system is morphodynamic (Huggett, 1980, 91–2). Consider population in an area: a city could be viewed as a 'deviation' of population density, caused by *cumulative causation* of city-developing forces of agglomeration and scale economies (see Table 7). Well-known expressions of this idea are 'Nothing succeeds like success', 'Unto every one that hath shall be given', and 'It's easy to make your second million'. *Negative feedback* opposes the main driving force leading to *homeostasis*, exhibited in an *ecosystem*, which is

a sort of superorganismic unity not alone between plants and animals to form biotic communities, but also between the biota and the environment. (Allee, 1934, 552)

. . . ecosystems—like all biological systems—exhibit homeostasis: that is, the ability to compensate for fluctuations in any part of their environment by negative feedback mechanisms. The more complex the ecosystem, the greater the homeostatic potential. It is, therefore, much more unlikely that catastrophic fluctuations in productivity will occur in a well-established tropical mainland ecosystem than in the environment of a newly-emerged arctic island. (Complexity of ecosystems increases, generally, with annual radiation total, age of ecosystem, and size of the area on which it is found.) (Chorley and Kennedy, 1971, 331)

Chorley (1973, 160–1; 1987, 381) has pointed out the difficulty of extrapolating this term into *human ecosystems* (cf. human ecology): 'In short, geographers are being faced with the basic problem of modelling systems which are stable in the short term under negative-feedback mechanisms, yet are capable of long-term changes under the positive-feedback mechanisms involved in economic and social tendencies' (1973, 161). *Goal-seeking feedback* is where man intervenes via *valves* or *regulators* to convert *process-response* systems to *control systems* in physical geography (see Chorley and Kennedy, 1971, via their index). A counterpart in human geography is goal-oriented planning (master planning), a type of socioeconomic planning that has often been superseded by problemoriented planning (see Bird, 1977a, Table 25, 138). A *leading part* of a system is a component, knowledge of which enables a description and prediction of outputs from the system and of the whole question of degree of centralization and decentralization in the political geography of a state. Ashby (1956, 202–18) proved that, where there is a regulating system and a system to be regulated, the former, to be competent in regulation, must be able to generate as many different states as the latter: 'This is the *Law of Requisite Variety*. To put it more picturesquely: only variety in R [the regulating system] can force down the variety in D [the dependent system]; only variety can destroy variety' (ch. 11, s. 1.7; my italic).

The *connectivity* of systems can be precisely indexed through the techniques of graph theory and topology (Kansky, 1963, developing the work of Garrison, 1960). The ability to assess networks quantitatively led Haggett and Chorley (1969, p. v) to 'argue that there are equivalent spatial structures common to both fields' of physical and human geography, an initiative which did not gain wide support. Several quantitative formulae can be used once information has been converted into nodes (vertices) and paths (edges), much like the schematic style of the London Underground map, with its stations and intervening lines.

In simple terms *entropy* is a state of maximum disorder in a system. Senior (1979, 206) has distinguished three meanings of 'entropy' that geographers may encounter:

1. A measure of uncertainty (Webber, 1972, 1977).
2. A descriptive statistic. (Both these interpretations are statistical).
3. A measure of increasing disorder linked to the second law of thermodyamics—a physical interpretation.

(The relationship between statistical and physical interpretations of entropy are explained in Cesario, 1975, 47–8). A.G. Wilson and his research team at Leeds have produced a family of entropy-maximizing models, in the statistical sense of entropy, as Hepple has explained:

Any specific trip distribution pattern . . . known as a 'macrostate' . . . can arise from many different sets of individual commuting movements or 'microstates'. Entropy measures the number of different microstates that can give rise to a particular macrostate . . . In the absence of detailed microstate data, we assume that each microstate is equally probable [the 'entropy-maximizing'], and that the macrostate with the maximum entropy value is the most probable or most likely overall pattern. (Hepple, 1981*a* 102; see also Wilson, 1981, 54 ff.)

Biotic systems feed on their task environment drawing *negentropy* (negative entropy) from it to sustain cellular specialization, until death supervenes when the Grim Reaper arrives in the shape of the second law of thermodynamics. Increasing order of information is an example of negentropic trends but only in so far as the data are capable of being understood by minds in the noosphere—the final system in the whole series of 'spheres' or systems. Added to the barysphere, lithosphere, hydrosphere, atmosphere, and biosphere, is the noosphere:

The recognition and isolation of a new era in evolution, the era of noogenesis [the genesis of mind], obliges us to distinguish correlatively a support proportionate to the operation—that is to say, yet another membrane in the majestic assembly of telluric layers . . . Much more coherent and just as extensive as any preceding layer, it is really a new layer, the 'thinking layer', which, since its germination at the end of the Tertiary period, has spread over and above the world of plants and animals. In other words, outside and above the biosphere there is the noosphere. (Teilhard de Chardin, 1959, 182; see also, Bird, 1963; Trusov, 1969)

Catastrophe theory embodies a series of mathematical formulations to account for the fact that systems may suddenly veer from one state to another quite different state as a result of a nevertheless steady change of input which causes the system to be moved across some critical *threshold*. Homely examples are 'the last straw that breaks the camel's back' and 'a watched pot does boil'. The sudden onset of spring in continental climates (the 'crack' illustrated musically in Stravinsky's *Rite of Spring*) is partially due to the fact that the sun gradually melts the snow cover, climbing higher each day; then, overnight the snow cover disappears and the sun strikes the receptive earth, which quickly warms the air above it. In a snow-bound Ottawa bus-shelter in late March, I once said to a fellow traveller: 'Doesn't spring ever come in this country?' The reply was, 'Say, bud, we don't have no spring. Here winter just flips over!' More generally, but still climatically, observe how a steady change in the sun's declination produces equinoctial storms.

Equifinality was defined by von Bertalanffy, originator of GST, as obtaining when the same final *state* (condition of a system at one time) may be achieved from different initial conditions and in different ways (e.g. peneplains in physical geography; and metropolitanization in human geography, see Bird, 1977*a*, Fig. 21, 128). Haines-Young and Petch (1983) note that equifinality has been used in geomorphology to state that the same end-products can be produced from different initial conditions by different processes, and they cite the example of arroyo formation. In such cases Haines-Young and Petch believe we should ask ourselves if this really obtains; or maybe our understanding of processes may be deficient—what are apparently dissimilar mechanisms are in fact the same; or final states that appear similar in fact may be different and in that sense reflect their diverse origins. They suggest that equifinality should be reserved for cases where the same final state is produced by the same processes from a range of initial conditions. This would preserve such laws as we make about causal processes. Earlier they had made the point that multiple working hypotheses are used to obtain a single one that in the end can be shown through testing to be better than the others; and Gerrard (1984, 365) agrees that equifinality should be accepted only as a last resort when all other hypotheses are rejected. Haines-Young and Petch's final comment in 1984 was that the concept of equifinality promotes complacency, and we have to devise tests to distinguish between our working hypotheses, which *prima facie* all seem to produce the same result. This discussion is a neat link between the systems approach and scientific methodology.

Systems in geography: advocacy

K.J. Gregory (1985, 143–4) identifies Chorley's 1962 paper as being very influential in moving geomorphology from an approach based on historical development to one using systems methods; and later in 1971 came Chorley and Kennedy's *Physical geography: a systems approach*—'enormously successful and must rank as one of the most cited and most influential physical geography textbooks of the twentieth century' (Gregory, 1985, 145). If we add Stoddart's (1967) chapter, Chapman's (1977) volume, and Bennett and Chorley's *Environmental systems: philosophy, analysis and control* (1978), the Cambridge grip on systems advocacy seems strong, though it is curious to observe that Chisholm (1967) was a critic of GST in geography long before he arrived in Cambridge, and Kennedy (1979) became another systems critic after she had left. Stoddart (1967) listed four advantages of the ecosystem concept: it is monistic (holistic), supplanting the dualism of environment on the one hand, and man, animals, and plants on the other; it enables structures to be investigated; it functions; and it is a type of general system, enabling geography to subscribe to geosystems and GST, thereby providing another argument against geography as somehow 'exceptionalist' to the general methods of science (cf. Schaefer, 1953).

A major area of debate has been the ability of systems theory to face so-called 'soft' systems (see Table 25), and, in addition, the 'systems in geography' movement has had to cope with the guilt-by-association technique of critics anxious to include it under the general backlash against positivistic methods. This has made some systems advocates a little cautious:

I do not think that the concept of system will have any great operational consequences for a long time yet. It represents an ideal that the real world does not fully approach. (Chapman, 1977, 6)

Despite its promise in geography, systems analysis has not yet achieved

TABLE 25.   The characteristics of hard and soft systems

Hard systems	Soft systems
Well-defined goals	Objectives frequently poorly defined
Well-defined boundaries	Boundaries poorly defined
Clearly established procedures	Decision-taking procedures vague
Quantifiable performance	Difficult to quantify
Clearly structured	Poorly structured
Physical systems	Human activity systems?

Note: Morgan (1981, 223 n. 9) points out that Bennett and Chorley (1978) use a different classification: ' "hard system" = concrete or physical; and "soft system" = abstract or mental' (see Fig. 9 above). In Checkland's (1981, 14) fourfold typology of systems—natural, designed physical, designed abstract, and human activity systems—only the last are soft systems.

Source: Agnew (1984, Table 1, 168).

anything approaching the operational status its intrinsic merit would seem to warrant. Held by the few geographers as a recipe for great theoretical and methodological advances, balked at by the many, its direct value to geographical practitioners has yet to be convincingly demonstrated. (Huggett, 1980, 24)

Huggett then immediately goes on to indicate the main charges against systems, one of the very rare instances of an advocate of a concept or a methodological argument acutally producing counter-arguments to his thesis and debating them. His very last words are modest:

What harm can come from using systems analysis in geography? Probably none. On the contrary, as this book has endeavoured to show, systems analysis is a useful implement which has already produced interesting results and turned up some unexpected findings. Surely, if geographers eschew the system-mongers, they will forgo in systems analysis a sophisticated and powerful method of study. (ibid. 193)

The last words of Bennett and Chorley (1978) are not quite as confident as you might have expected. After emphasizing a 'truth which has emerged from this book'—that environmental systems are complex and that efforts to control or influence these systems have been made with reference to aims 'held only by small, articulate and influential groups of society', they conclude thus:

This book has no easy answer for this ultimate control problem except to exhort an increasing realism* through systems modelling. Three matters are of particular concern; firstly, the aims of control and intervention, secondly, the results of such control (unplanned as well as planned), and thirdly, to what degree the aims of control and the intensity and hierarchical level of control must be moderated in order to result in broadly benign effects. On the one hand, it may be that it is merely the inadequacies of current systems conceptions and control techniques which separate a Kafkaesque hell from a Marxist heaven; on the other hand, it is much more likely that all but the most broadly humane and delicate of controls over environmental systems may be disastrous. (p. 553)
[*Careful! Remember the date is 1978, so this is not transcendental realism.]

Observe that the word 'control' appears seven times in this passage, and also again in the very title of the book. At this point humanist geographers can hardly be restrained from asking: 'Control by whom? For whom? For what purposes? And how is all that to be decided? And by whom?'

Struggling to answer such questions, Morgan (1981) and Agnew (1984) have drawn attention to Checkland's (1972) approach to soft systems. This involves attempting to formulate a 'root definition' of a system at an early stage, a statement of 'the function or purpose of one system thought to be most relevant to the problem' (Morgan, 1981, 222; see also Table 26). An example that Huggett (1981, 224) quotes from Checkland is the various possible root definitions for a church: 'a social welfare system, a ritual-organising system, a belief-maintenance system, or a system to provide support in the face of unanswerable questions'. What we appear to have is a number of possible systems operating in the same space–time framework, and our choice of which to study depends on the problem we wish to solve. Huggett (1981, 225) seems to agree with this, though he is doubtful about aspects of the Checkland approach: 'The selection of variables is up to the reseacher and is thus a subjective process' (see Table 26).

If GST invites importation of general methods into geography, it is also possible to move systems thinking from one area to another *within* geography: 'We conclude that the hierarchical nature of intra-urban retail structure may be attributed to the same process that generates systems of central places. The primary difference is the high density of households within cities' (Berry

TABLE 26.   Checkland's mnemonic for six crucial characteristics of a
root definition within human activity systems (CATWOE)

C Customers	Beneficiaries or victims of a notional system
A Actors	Persons or agency who cause to be carried out the activities of the system, especially its main transformation
T Transformation	Core of the definition, how the system's inputs are transformed into the outputs
W *Weltanschauung*	Outlook, framework, perspective, world-view, which gives meaning to the particular root definition
O Owners	Persons or agency with prime concern for the sytem and power over its 'life and death'
E Environmental constraints	Impositions from the system's environment or from wider systems of which it forms part which must be taken as given

Source: Checkland (1981, 18 and 224–5, and glossary, 312–19).

and Horton, 1970, 456). The germination of the idea of the necessity for three different systems within cities arose in Berry's (1959) paper, where he first mentions the idea (also citing Bunge, 1958), and then it blossoms into a typology in 1963.

Two other advocates of systems in geography are Wilson (1981) and Warntz (1973). Wilson's entropy-maximizing approach has already been mentioned, and in a general book of 1981 the following appears in an opening section on aims:

The usual emphases of a systems-analytical study . . . restrict the field of application to some extent: the main concern is with *complicated* systems whose components exhibit high degrees of *interdependence*. [Wilson, 39 ff. makes explicit use of Weaver's (1967) classification and is mainly interested in Weaver's systems of organized complexity.] The behaviour of the 'whole' system is then usually more than the sum of the parts. (p. 3)

The lineage of Warntz's (1973) argument is contained in his title: 'New geography as general spatial systems theory—old social physics writ large?' As its name implies, social physics draws parallels between the physical and social sciences and is associated with its founder, J. Q. Stewart, of the Department of Astrophysical Sciences, Princeton. Warntz (1984) has described how he became a collaborator with Stewart:

I came upon a copy of *Coasts, Waves and Weather* [*for Navigators*] by Stewart [1945]. This book had been prepared primarily to explain to marine and air navigators the physical environment in which navigation must be carried on. Fortunately for me, Stewart could not resist the temptation to include an exotic chapter describing potential of population and its sociological importance . . . demonstrating that physics's double contribution consisted not only of mechanical technology but also—and perhaps more importantly—patterns of thought. (p. 141)

The thesis of Stewart and Warntz (1958) is that at a sufficient level of abstraction (macrogeography) the cross-referencing of systems with different substantive correlates is possible—the very essence of GST, and hence general spatial systems theory. A founding tenet was the gravity model, which was extended from a system dealing with relationships between two points to a system mapping the influence of all other points on each one to produce maps of potential—most commonly population potential.

As we perceive molecules and people in well-defined circumstances to behave as aggregates, we will be satisfied to use common functional cause–effect relationships. That this structure—these relationships—is the set of abstractions by which we originally understood the physical behaviour of a gas does not limit it phenomenologically.* On the contrary, the fundamental postulate here is that structures, saying whatever they do about our perceptions, may be quite general. (Warntz, 1973, 91) [*An adverb derived from 'phenomenon', and not the epistemology called phenomenology.]

Your own particular reactions to that passage will say a lot about your philosophical position as a geographer.

A more recent study based on general spatial systems is by Coffey (1981, 239), exploring 'the question of whether or not a certain (small) number of systematic spatial properties are evident in phenomena that may be judged to differ widely in their non-spatial aspects'.

At interdisciplinary level, sophisticated computational technology fuels ambitious research programmes, which enlist systems thinking. The following are currently being supported by scientists funded by the British Natural Environment Research Council: Global Atmospheric Research Programme (GARP), the North Sea Programme, and the World Ocean Circulation Experiment (WOCE).

Systems in geography: criticisms

For convenience, the criticisms that systems have attracted, mainly from geographers, can be divided into 'fundamental–conceptual' ('Should I use GST or systems at all?'), and operational ('Systems have these problems when you try to use them'), and these latter objections are summarized in Table 27.

TABLE 27.   Operational problems involved in applying systems methods within geography

---

1. The problem of remembering, while systems-building, systems operating, and contemplating systems results, that the system is a concept and not necessarily an actual system in the 'real world' (the problem of reification).[a]

2. Systems thinking channels problem-solving into one scale of approach, possibly losing profitable trade-offs between different scales of approach. More specifically, GST channels problem-solving exclusively into macroscale approaches.

3. All geographical systems are open, and this results in a problem of where to apply closure, or what is the system and what is its 'environment'. 'Solution of this problem, a process called entitation, is doubly difficult in geography because closure must be effected in two explicit considered dimensions, the spatial and the "phenomenal" ' (Langton, 1972, 133). (But see Chapman's comments on this, 1977, 33.)

4. Conversion of geographical data into systems-type elements and relationships usually requires major assumptions.

5. The act of systems-conception may block off possible non-system types of explanation; systems are seen everywhere; and the problem may reside in the relations of a single phenomenon to its environment.

6. Novelty can be injected into a system as it is allowed to run diachronically[b] into the future, but systems thinking *per se* does not necessarily generate the novelty which may be conceived in an area outside the system. There is the allied difficulty of measuring the impact of the novelty where feedback loops are in operation within the system.

7. GST asserts that systems thinking applies analogically in any discipline, but in reality one has to pay attention to the constraints of the

substantive data. (The same caveat applies to so-called 'general spatial laws'; see Sack, 1973, 1978.[c] The 'gravity' analogy can be avoided by expressing the gravity model in the form of a mathematical equation.)

8. 'Systems' is a term used loosely, and, while GST promotes a unified approach in all disciplines, each discipline has a slightly different consensual concept of 'system' coloured by its own substantive data. This can cause confusion when translating from other disciplines into geography or when geographers participate in interdisciplinary studies.

9. Impact of external stimuli on a system is subject to different lags (when feedback loops are operating between subsystems). '. . . the greater the complexity of these external links, the greater, too, is the probability of response. Variety of linkages is thus a major determinant of the degree of response that is possible to a stimulus: the inertia of a system is inversely related to the degree of its external connections' (Langton, 1972, 140).

10. It is much more difficult to model images of phenomena compared with modelling phenomena themselves.

11. Systems are not so efficient where decision-making is uncertain, emotional, or qualitative, or where there is a strong probability of novelty irruption during the life of the system. 'In the comparison between human and physical systems, a critical distinction arises in that the former are self-conscious and generally considered to be goal-seeking while the latter operate according to the laws of physics' (Coffey, 1981, 241).

12. In human spatial systems, general models which include all relevant variables 'are so complex that our present knowledge is inadequate to construct them successfully' (Klaassen, Paelinck, and Wagenaar, 1979, 157).

13. The problem of disaggregating the too complex general models into submodels without losing a principal attribute of general models—the interrelationships between the variables. 'The first and general principle should be . . . to cut the sub-models out of the general model in such a way that a minimum of interaction flows are interrupted' (ibid.).

14. Regional systems models, based on regression from past values and also on spatial diffusion effects from outside the region are still 'black box' models (where mathematical and systems relationships are posited without causal laws being understood). In such black box examples, systems thinking may paper over gaps in knowledge of cause and effect sequences.

15. Systems thinking is functional thinking, making a virtue of using general invariant laws wherever possible as useful analogies; but human societies contain important elements and relationships which are historically specific (historicity, configuration, and *conjoncture*, see text).

TABLE 27. Continued

---

16. The problem of goal definition in human activity systems. In cybernertic goal-seeking-with-feedback systems, the goal is provided from outside the system. But in human activity systems the goal may be the appreciation of the ongoing system relationships, which conditions changes in these relationships and is also modified by the changes.[d]

---

[a] Checkland (1981) flirts with reification. He has four classes of system—natural, designed physical, designed abstract, and human activity systems. He says the first two 'could not be other than they are' (p. 14) and the last is 'less tangible' than natural and designed systems (pp. 110–11). My argument is as follows: a car is a 'designed physical system'; a car exists as a tangible phenomenal object, with tangible phenomenal parts; but it (and also 'natural systems') exists as a *system* only when a mind understands it as such.

[b] Langton supplies a pithy definition of 'diachronic': 'The distinction recognised in geography by Darby [1962b, 132–4] between what he termed 'cross-sectional' and 'vertical theme' studies, is generalised by systems theorists as being one between "structural" or "synchronic" and "diachronic" analysis' (Langton, 1972, 136).

[c]Smalley and Vita-Finzi (1969) conclude that the application of GST to earth sciences 'lends confusion rather than clarification' (p. 1593).

[d]A paraphrase of Checkland (1981, 262–3), and partially derived from Vickers (1970).

Source: Much of this table is derived from Chisholm (1967), Langton (1972), and Huggett (1980, 24–5), who says that 'The main factor [of operational difficulty] seems to be the sheer complexity and structural richness of systems of interest to geographers, which are replete with non-linear relations, complicated by lags, thresholds, and a propensity to lurch from one state of disequilibrium to another', and here he cites Chorley (1973, 164).

As we have seen, the most potent advocacy for GST in geography derived from the social physics initiative, notably Stewart and Warntz (1958). Warntz (1973) summed up their position as follows:

In 1958 John Q. Stewart and this author identified a 'macrogeography' of human phenomena including the recognition of number of people, distance and time as basic categories or dimensions. It was argued that only from the 'thorough appreciation of these brute physical factors can grow a correct and fertile treatment of the loftier human characteristics, which, too, must be included in their turn'. (p. 90)

As Johnston (1979b, republished 1987, 72–4) points out, Lukermann (1958) was an early objector to this initiative of general spatial systems theory. He bluntly stated (p. 5) that 'explanation in science is never explanation of observed reality, but only the explanation of the consequences of hypotheses'. The test was not whether one could show isomorphism between mechanical models and models of social behaviour but via comparisons between these latter models and observed empirical reality.

One can see that for Stewart and Warntz 'brute physical factors' or 'certain laws which govern all physical and human phenomena' (Coffey, 1981, 4) have primacy. This has been called functionalism by certain human geographers (notably D. Gregory, 1980), drawing attention to developments in social theory, with a critique based on functionalism's inadequate recognition of the historicity of societies (each age should be interpreted in terms of its own social objectives and principles); and GST also places inadequate emphasis on the role of human agency. A somewhat similar objection has been made by Kennedy (1979). She drew attention to the words of palaeontologist Simpson (1963) with his summary of 'two, complementary aspects of scientific explanation as the *configurational* (i.e. that relating to and/or determined by unique conditions of time and space) and the *immanent* (representing 'unchanging properties of matter and energy and the likewise unchanging processes and principles arising therefrom', Simpson, p. 24)' (Kennedy, 1979, 552). A similar duplex perspective on time has been expressed by Braudel (1972, 353) as *structure* (Fr., long-term, 'permanent', slow moving) and *conjoncture* (Fr., short-term, ephemeral, fast-moving). Obviously, systems are more successful with the first than with the second, or than with a combination of the two.

A more general objection to initiatives in human geography derived from GST is based on the mechanistic view of society that is necessarily taken (Gould, 1981, 312). More specifically, Gregory (1980) argues that the dependency of systems theory on an operational technology is also an advocacy of a particular ideology:

Thus, as Scott and Roweis (1977, 1099) have convincingly demonstrated in the specific context of urban planning, systems theory has an essential *historicity* which is characteristically obscured: it 'abstracts away from real, historically determinate parameters of human activity, and gratuitously assumes the existence of transcendent operational norms', whereas 'what motivates action is human *interests*' which although they may 'take on the ideological *appearance* of norms . . . are primarily ensembles of concrete practices, and only secondarily systems of abstract ideas or values' (p. 332)

Gregory (1980, 341) ends by quoting Marcuse (1972, 130) who observed how 'today domination perpetuates and extends itself not only through technology but as technology, and the latter provides the great legitimation of the expanding political power'.

Systems in geography: scrutinizing the moving balance sheet

The foregoing survey clearly shows that systems have found a readier acceptance in physcial geography compared with human geography, notwithstanding the advocacy of a general spatial systems theory. Seven of the 'operational problems' in Table 27 occur only in human geography, and most of the criticisms of the approach have come from human geographers. Langton (1972, 170) believed that systems thinking is essentially an empirical methodology which facilitates the conceptualizing of behaviour and is a route to causal statements about that behaviour. Then comes this prescient sentence: 'This empirical bias is inevitable given the fact that systems theory is concerned with diachronic analysis' (see Table 27 n.b). Events since 1972 have demostrated that as geography, both physical and human, has focused more and more on patterns of change, systems theory has come more and more into use. The balance sheet of advantages and disadvantages of systems theory seems to have remained in the black, but, to go beyond a litany of problems and criticisms, perhaps the following five points need bearing in mind:

1. Systems theory provides *a* methodology for geography, but not *the* methodology.

2. For the foreseeable future, systems theory is likely to be of relatively more importance in physical geography and in those areas of human geography where cause and effect sequences are in operation at a sufficient scale of generality (macro-approaches).

3. Systems can of course be used elsewhere in human geography, provided that they are used as an aid, and their results not taken as providing normative guide-lines.

4. Systems theory will become more sophisticated and be able to deal with problems thought intractable for it at present, such as detailed analysis of energy flows between components, incorporation of novelty, goal-seeking aspects of behaviour, and even humanistic features such as the sympathetic understanding of human actors within systems. This explains the suggested

increased size of the area representing 'systems theory solutions' in the future in 'T_3' of Fig. 19; but observe that the area of problems 'solved' by other methods also increases as does the number of problems recognized as awaiting solution (symbolized by the larger circumference of the circle).

In physical geography K. J. Gregory (1987, 3) shows how a new focus on energy flows links up with more 'dynamical' systems:

. . . the major achievement of the systems approach to date has been to describe the structure of systems in nature rather than focusing sufficiently upon the energy flows which link the structural components and which therefore depend upon the power distribution in particular systems. The possible integrating role of the power theme can also apply to the other themes recently evident in physical geography. Thus studies concerned with change over time basically depend upon the ways in which power expenditure in nature varies over time and upon the way in which such changes can be dated, modelled and explained.

Systems 'concerned with change over time' are the theme of Thornes (1987). He demonstrates advances over an equilibrium approach to systems 'which largely ignored the ways in which the equilibrium had been reached through time' (p. 28). The adjective 'dynamic' was certainly used in the early days of systems in physical geography, but dynamic systems theory merely stressed the functional aspects of a working system. Thornes shows how *dynamical* systems theory explores behaviour of a system through time (diachronically). He demonstrates (Fig. 1.2.3,33) the ways in which 'state variables' of a system behave over time: damped behaviour (homeostasis evident); explosive behaviour (homeorhesis—'the last straw cometh'); periodic behaviour (as in a 'continental' climate); or unsystematic behaviour; and there is a richness in the study of the mixes of these behaviours by any one system over time.

5. In Fig. 19 there is a 'crescent of subjective judgement' necessary when using systems, particularly in areas of human geography. This needs a simple illustration. In a large organization, the financial division must use all the sophisticated techniques of accounting to provide analysis and projections. Let us suppose it makes use of a systems model which runs into the future to provide an output—a financial forecast. The management team of the organization must study this methodology and look particularly at any counter-intuitive elements thrown up by the model. There

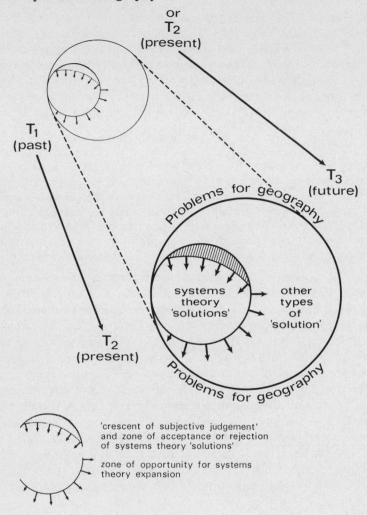

FIG. 19. *Systems theory solutions: one methodology among others in geography*

Note: For 'crescent of subjective judgement' see text. Notice the allometric growth over time from T_1 to T_2. The law of allometric growth states that the growth of a subsystem is proportional to the growth of the system as a whole. This is to suggest that systems-type 'solutions' will grow, but so will other types of proffered 'solutions', and so will the problems faced by geographers, symbolized by the expanded circumference in T_2. The sizes of the smaller circles within the larger would be proportionally greater for physical geography considered by itself, and proportionally smaller for human geography alone.

must now enter a subjective judgement as to whether these elements will have an actual counterpart in the future, or whether they are partly or wholly derived from the assumptions on which the systems model is based. The management then has to take decisions of this 'subjectively judged' output of the model, and do what this modified output suggests, not literally, but bearing in mind (a) the original assumptions, (b) the objectives of the organization (in geography, these are the attempts to solve problems); and (c) the possible novelties (including acceptable compromises) that can be injected into the future operation of the organization.

My own 'crescent of subjective judgement' about systems theory is that the balance of advantage is such that it will have an important part to play in geography, with relatively more emphasis in physical geography. The word 'crescent' is appropriate because subjective judgement will have to grow at least in line with systems sophistication. We shall have to think of 'systems' as embracing models operated and thrown up by computers, AI (artificial intelligence), and whatever exosomatic (outside our minds and bodies) thinking systems are there to aid us. But, however sophisticated these aids become, the crescent of subjective judgement will be needed as a catalyst and also as a protective membrane between us and the technical apparatus we have invented as in combination we confront the problems identified. Otherwise the phenomenal and behavioural environments that we model with our systems will cease to be a human world.

8

Geographical Imagination and Three Worlds

Types of imagination and creativity

Methodological discussion of such basic topics as systems theory, ways of 'doing science', and the sheer getting of ideas leads into interdisciplinary areas. Just as there is no consensus on the definition of the 'scientific method', there is no received view of creativity, or of the basic constituents making up the creative act. Yet, as geographers, we wish to make our subject richer in ideas pregnant with other ideas in cumulative fashion. In this we have aims in common with workers in all other academic disciplines, and there is a whole literature on theories of creativity. The first thing to establish is the type of imagination we are prepared to bring to bear on our problems. If we know very little about a subject, we must submissively learn the rules of the particular game; later we can add to the richness of our learning experience by looking critically at the rules. In geography, unlike some other subjects, this can happen in the undergraduate curriculum. By contrast, I once heard a Professor of Biochemistry say in an inaugural lecture that his first-degree students had no possibility of contributing to discussion of the subject since they were engaged in the full-time job of assimilating what was already in intellectual place.

Bartlett (1928) describes three types of imagination (see Table 28), and, while the beginner is likely to employ Type I thinking, he is soon encouraged to progress to Type II. The professional geographer with a proven methodology and a well-established epistemology will no doubt more often employ a Type III imagination, but Bartlett points out that constructive imagination may be interspersed with periods of assimilative imagination and creative interpretation; and these may be the most exciting parts of the work (p. 84). The table may strike home to my readers; I have done my best to warn against remaining in a Type I submissive reaction. In a broad sense Bartlett's three types of imagination

TABLE 28. Types of imagination

Type of imagination	Process involved	Critical faculty employed	Temperamental attitude
I Assimilative	Focusing on topic; attempting understanding; 'resonance of feeling' with objects of study	Absent	Capacity for wonder; rather submissive (in extreme cases, proneness to brainwashing)[a]
II Creative interpretation	All above are present, but individual does not quite lose himself; existing material to be interpreted	Criticism present, but not related to a perceived datum in object of study but based on own past experiences	Capacity for wonder; sympathetic co-operation with object of study
III Constructive imagination	No resonance of feeling, but imagination characterized by a purpose; material collected in light of a plan which may undergo changes (sorting and criticism); existing material seen as a problem	Critical analysis dominant	Wonder remains, but attitude of dominance or mastery over material via a 'certain uncompromising aggressiveness' = a type of imagination most likely to arouse opposition

Note: I believe that at a sufficient level of generality, similar types of imagination and creativity are to be found in both arts and sciences, so that in this context it seems to matter little that geography covers a spectrum which at one pole is close to the natural science model and at the other close to methods of the humanities. However, we may note Bartlett's gloss about 'constructive imagination': in artistic thinking the 'broad strategy . . . is neither to compel by proof nor combine by assertiveness; but to satisfy by attainment . . . the requirements of a standard' (1958, 197).

[a] Material in brackets added.

Source: Derived from the text in Bartlett (1928).

correspond respectively to the standards required at school, in first-degree study, and at research level, although the undergraduate surely grasps what is meant by constructive imagination since he is exposed to so much of its fruits. At all levels, a form of creativity is required, and there are plenty of studies to suggest *what* this is. Three examples are Wallas (1926), Koestler (1964), and Arieti (1976).

Wallas proposed four stages in creative thought: preparation, incubation, illumination, and verification. Obviously, the act of creation is represented by stage 3 and has been prepared somehow though not fully specified, in stages 1 and 2.

I find it convenient to use the term 'Intimation' for that moment in the Illumination stage when our fringe-consciousness of an association-train is in the state of rising consciousness which indicates that the fully conscious flash of success is coming. . . . if this feeling of Intimation lasts for an appreciable time, and is either sufficiently conscious, or can by an effort of attention be made sufficiently conscious, it is obvious that our will can be brought directly to bear on it. (Wallas, 1926, 97 and 99)

Koestler proposed the term 'bisociation' for the act of creation, when two contrasted frames of reference or 'matrices of thought' are brought together to form a sudden and productive synthetic fusion. Again, we find Koestler unspecific on the stimulation of the creative act, on how we identify the frames of reference that are going to be so productively combined. Lombardo (1987, 360) had a comment on this difficulty. '. . . one can never totally explain originality and creative insight or else it would not possess novelty. Something out of the ordinary must occur, something unexpected, and then the unusual must be seized upon and perceived as an opportunity for growth rather than being a nuisance,' (Lombardo, 1987, 360). Arieti starts from a base in psychiatry: creativity occurs when primitive forms of cognition (the primary process function of the unconscious part of the psyche) are appropriately matched with secondary process mechanisms of conscious common logic.

I have proposed the expression *tertiary process* to designate this special combination of primary and secondary process mechanisms. (Arieti, 1976, 12)
. . . the primary process (or poetic, metaphoric, mystic, primitive, archaic, childlike)

Our conscious intellect is too exclusively analytic, rational, numerical, atomistic, conceptual, and so it misses a great deal of reality, especially within ourselves. (Maslow, 1967, 53)

The unconscious might be described as containing what we know when we are not thinking about it. 'Preparation' and 'incubation' in Wallas's sequence are where we enrich the unconscious, and there are some tricks of the trade to lubricate this rather shadowy area of brain processes to keep in creative trim. I have tabulated a few of these tips from an appendix 'On intellectual craftsmanship' from C. W. Mills's (1959) *The sociological imagination* (Table 29); and a counterpart for the 'harder' areas of geography, closer to the natural science model, is P. B. Medawar's (1979) *Advice to a young scientist.*

TABLE 29.   Mills's advice on creativity

Incubation–preparation
1. Set up a file. Put in it some good phrases, ideas in the literature, and also your own ideas. 'By keeping an adequate file and thus developing self-reflective habits you keep your inner world awake' (p. 197).
2. Make written plans of what you are proposing to do, and periodically review them.
[Additional note: if you date the plan and put it away, when you come back to it some time later, you will be able to view it with a fresh perspective which will foster self-criticism, and this will often take the form of a new idea.]
3. Rearrange the filing system. In this process 'you often find that you are . . . loosening your imagination' (p. 201). 'Imagination is often successfully invited by putting together hitherto isolated items, by finding unsuspected connections' (p. 201).

Confronting problems: four stages (p. 206)
'Facts discipline reason; but reason is the advance guard in any field of learning' (p. 205).
1. Distinguish elements and definitions which from your general awareness are concerned with your topic, issue, or problem.
2. Attempt to see the relationships between these definitions and elements 'building . . . little preliminary models' (p. 206).
3. Eliminate error due to omission of necessary elements, unclear definitions, or wrong emphases.
4. 'Statement and re-statement of the questions of fact that remain' (p. 206).

TABLE 29.   Continued

Stimulating the imagination: seven methods

'. . . essence [of imagination] is the combination of ideas that no one expected were combinable' . . . (p. 211).

1. Rearrange the file, being passively receptive to novel linkages.

2. Attitude of playfulness towards phrases and words, so as to choose exactly defining words from among a choice. Move up and down a hierarchical scale of meaning.

3. Reclassify. 'A new classification is the usual beginning of fruitful developments' (p. 213).

4. 'Often you get the best insights by considering extremes—by thinking of the opposite of that with which you are directly concerned' (p. 213).

[Additional note: if you construct a continuum for your data, or problem facts, with two poles, you can compare each of these, or any point on the continuum with either extreme. Such a procedure may throw up interesting comparisons which may indicate significant points on the continuum.]

5. Temporarily invent your own scale. If things were at a different scale, would that make a difference?

6. 'Whatever the problem with which you are concerned, you will find it helpful to try to get a *comparative* grip on the materials. The search for comparable cases, either in one civilization and historical period or in several, gives you leads' (p. 215).

7. Distinguish between topics and themes. 'Usually most of what you have to say about a topic can be put into one chapter or a section of a chapter' (p. 216). A theme is a more general idea, some 'master conception' or a 'key distinction'. May be more than one theme. '. . . you will ask of each topic: Just how is it affected by each of these themes? And again: Just what is the meaning, if any, for each of these themes of each of the topics?' (p. 216)

[Additional note: Mills does not express himself too happily here. Here is my version. When you are dealing with or erecting a generalization, say to yourself: what case-study material exemplifies this? When poring over the detail of a case-study, say to yourself: what generalization does this material illuminate, confirm, or deny?]

Source: This table is derived from a 31-page appendix in C. W. Mills (1959, 195–226), 'On intellectual craftsmanship'. The page numbers refer to direct quotations from this appendix, whose purpose is 'to report in some detail how I go about my craft. This is necessarily a personal statement, but it is written with the hope that others, especially those beginning independent work, will make it less personal by the facts of their own experience' (p. 195).

The subheadings and the 'additional notes' are mine. Over the years I have reread Mills's appendix many times 'to get in the mood'.

Getting ideas: the Moles classification, with geographical examples

In discussions of creativity, one cannot help observing how a form of comparison emerges as a key methodological operation.

The conditions for original thinking are when two or more streams begin to offer evidence that they may converge and so in some manner be combined.

. . . The most important of all conditions of originality in experimental thinking is a capacity to detect overlap and agreement between groups of facts and fields of study which have not before been effectively combined, and to bring these groups into experimental contact. (Bartlett, 1958, 136 and 161; this and other work on creativity appear in an anthology compiled by Vernon, 1970)

Once the importance of comparison is acknowledged, one can then see a similarity of method in science, literature, and humour (see Table 30). An important corollary is that such a correspondence

TABLE 30.   Methodological mechanisms involving 'comparison' (by one in a position to compare the methodologies of arts and science—a geographer)

Mainly science	*Technical terms in literature*	*Humour*	*In particular fields of study*
Analogue theory	Alliteration	*Double entendre*	Adaptation level
Comparison (comparing x with y; comparing x to y)[a]	Ambiguity[e]	Joke (via bisociation)[g]	theory[h]
	Analogy	Pay-off	Adoptive/adaptive
	Antithesis	Play on words	dichotomy[i]
Constant-revision (of our knowledge of nature)[b]	Chiasmus	Pun	Ambiguity[j]
	Cliché (worn-out comparison)	Punch line	Association[k]
Correlation		Satire	Bringing back
Diagonalism	Comparison (both senses)	Take-off	(ideas for comparative purposes)[l]
Equations			
Exponents	Figurativeness		
General Systems Theory	Homonymy		Castle and border[m]
	Irony		
Hypothetico-deductive system (making *before* matching)[c]	Litotes		Centre–periphery models
	Meiosis		
	Metalepsis		Cognitive dissonance[n]
	Metaphor[f]		
Interplay	Metonymy		Complementarity principle[o]
Isomorphism	Oxymoron		
Model-making	Parable		Dialectical relationship[p]
Regression	Rhyme		
Scientific (descriptive) laws[d]	Simile		Energy-tension[q]
	Syllepsis		Filtering[r]

TABLE 30.   Continued

Mainly science	Technical terms in literature	Humour	In particular fields of study
Simulation (models)	Syllogism		Generated and
Synchrony	Symbol		generating
Systems	Synecdoche		cognitive
Theory-laden observation	Synonymy		representations[s]
	Trope		*Gestalt* law of
	Zeugma		closure[t]
			Icon
			Interaction[u]
			Map
			Metonymy[v]
			Plurisignation[w]
			Polychronism[x]
			Projection[y]
			Prospect and
			refuge[z]
			Reciprocity
			(integrating the
			opposition
			between the self
			and others)[aa]
			Reflexive
			relation[bb]
			Relativity (space–
			time)
			Semiology[cc]
			Sfumato[dd]
			Sign, signal,
			signum[ee]
			Sprezzatura[ff]
			Symbiosis[gg]
			Symbol
			Synaesthesia[hh]
			Token
			Trade-off
			Verisimilitude[ii]

Notes: Only a selection of sources is given below—for less familiar terms or for technical use of familiar terms.

[a] Comparing x with y means instituting a comparison, whereas comparing x to y means placing in the same class as y. The comparative process, in the first sense, plays a key role in Laudan's theory of scientific growth, e.g. 'What matters is not, in some absolute sense, how effective or progressive a [research] tradition or theory is, but, rather, how its effectiveness or progressiveness compares with its competitors' (Laudan, 1977, 120).

[b] Rescher (1978, 51).

[c] Gombrich (1960), see index.

[d] 'Our "laws of nature" are in fact the product of mind–nature interaction' (Rescher, 1978, 165).

[e] In the poetic sense, see Empson (1953).

f For a general essay stressing the importance of metaphor in human geography, see Tuan (1979*b*): '. . . the belief that science and metaphors are talking about the same things in different terms and that each can help explain the other'. Park (1980, p. ix) and Livingstone and Harrison (1980, 128) also discuss the comparison between scientific models and metaphors, with supporting references.

g Koestler (1964, 35 ff.).

h Helson (1964).

i Alchian (1950), Tiebout (1957), and Krumme (1969).

j In the sense of 'duplexity of meaning' (Rapoport and Kantor, 1967, 210); also used by Gombrich (1960), 'the same song of this book' (p. 313).

k In cognitive psychology refers to mental connection between concepts, percepts, sense-data, and memories of these so that one may evoke the other in associative comparison.

l Maruyama (1963, 179); cf. (comparisons are contagious!) '. . . constructive thinking demands the bringing together of realms of interest which ordinarily, so far, have not been connected' Bartlett (1932, 313).

m Ardrey (1967, 169–70), relying on Darling (1952).

n Festinger (1957).

o Pattee (1978); Park (1980, 112 ff.); Jantsch (1980, 24 and 272–4)

p As in 'dialectical relationship between individual and environment' Golledge (1979, 116); for a general survey in geography see Marchand (1979). See also Thrift and Pred (1981): '. . . time-geography is fundamentally dialectical' (p. 283).

q A term used by Foss (1949, 57–60) to describe the metaphorical process.

r Mooij (1976): '. . . the process of filtering, projection or interaction . . . need not be absent from comparisons' (pp. 170–1).

s Occur in the 'syntaxic stage', most mature level of Sullivan's (1953, 25–30) three modal stages of experience.

t A figure which is incomplete is perceived as if it were complete.

u See nn. d and r.

v Important alongside metaphor in the structural analysis approach to social anthropology in the 'logic by which symbols are connected' (Leach, 1976, see Fig. 1, 12).

w Introduced by Wheelwright (1954) to extend the 'either–or' relation of Empson's ambiguity (see n. e above) to a 'both–and' plurisign.

x Hall (1966, 163) suggests that Americans prefer one thing at a time rather than mixing (comparing) several activities at once (polychronism); hence the comparison between the strung-out Main Street and the polychronic European squares, plazas, and piazzas—see also Price (1968, 59).

y See n. r.

z The theory behind the experience of landscape in Appleton (1975).

aa Lévi-Strauss (1969, 84).

bb D. Gregory (1978, Fig. 2, 58, 76); see also Tyler (1978, 460).

cc See n. v; and Barthes (1968, in translation).

dd Invented by Leonardo da Vinci. A deliberately blurred image or veiled form cuts down the information on a canvas and thereby stimulates the process of projection (based on Gombrich, 1960, 144–5).

ee See n. v. For 'sign' see also Tuan (1978*b*).

ff This is where the possibility occurs of projecting one's imagination into the brush-strokes of a picture to produce a pattern (Gombrich, 1960, 144–5).

gg '. . . what survives is not the strongest, but the most symbiotic' (Maruyama, 1973, 436).

hh Discussed as synesthesia by Tuan (1978*b*, 366).

ii Popper (1972, 47).

of method may be found in the purest of physical geography and the most humanistic of human geography. 'May be found' not 'will be found' because nobody is forced to use hypothetic-deductive methods, which rely so much on comparative mechanisms. Such mechanisms may be found in other methodological frameworks, but I have felt bold enough to provide a 'typical Ph.D. thesis layout' based on a structure which has common elements in Popperian science and Greek drama (Fig. 20). Surely this is not the only possible structure, but it is one that gets results in a reasonable length of time because very often it places the investigator in a no-lose situation. If the hypothesis is a comparison between 'what could be true and what is in fact the case' (Medawar, 1969, 59), and this turns out to be confirmed, happy is the researcher; if there are residual problems, he and his reader may learn even more from what we may term 'the problem of the residuals'. But I must stress that 'scientific method' is really a whole family of procedures.

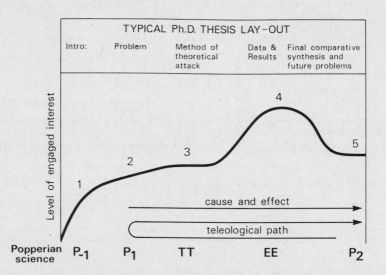

FIG. 20. *Comparisons between the structure of a Ph.D. thesis layout, two types of scientific inference, and classical drama*

Note: P_{-1} = problem orientation; P_1 = initial problem; TT = tentative theory; EE = evaluative error elimination; P_2 = remaining problems, different from P_1 (Popper, 1972, 199, with discussion; P_{-1} has been added).

The numbers are explained in the following passage by James Agate, drama critic (1928): 'The tragedies of Shakespeare, Racine and Corneille have five acts because though the Greek play had only one act the emotion of that act went through five distinct stages. There was first the beginning of the story, second its growth and complication, third a state of suspension or gathering clouds before the storm, fourth the climax or thundercrash, and fifth the clearing up.'

Source: Bird (1981, 142).

It is almost (but not quite) needless to observe that in speaking of 'the scientific method' of inquiry and thesis substantiation, we have no wish to deny the internal complexity of this methodology—that what is at issue is a diversified organon of methods and not one single all-enhancing method. The situation of mathematics provides an analogy. When one speaks of the probative method of mathematics as that of demonstrative proof procedures there is no wish to deny a pluralism of proof procedures . . . (Rescher, 1977, 177 n.)

We have seen that there are many attempts to describe the 'what' of the creative act. But of all the attempts to describe *how* to get ideas, perhaps the most comprehensive is by Moles (1957, 1964). This enterprise is Popperian in spirit as the following indicates:

Scientific knowledge establishes itself by a struggle, by a confrontation between the individual and the external world, and by a series of dialectical confrontations between what is at present established, forming the image of the *Universe here* and *now* and what the scientist is *going* to establish in opposition to that which exists. . . . In other words, because the building is never finished, there is no point in using syllogisms to plaster over the faults, the gaps, the errors in construction; but on the contrary the aim is to leave them well in evidence, to attract attention to them as an invitation for their correction. (1957, 13–15; translated from the French)

Moles's (1957) classification of twenty-one methods of scientific creation was later expanded by him to a mind-blowing forty-six methods (1964), but with the following concluding note: 'Differentiation between the methods is far from being rigorous. Many intersect with others where there are similar attitudes of thought but in different frameworks' (p. 81). Table 31 is an attempt to indicate these family linkages, if the list is restricted to the original 1957 formulation, which is given below. After the bare title of each method, a short note of explanation or expansion follows (based mainly on material in Moles, 1957); and then examples are given from the field of geography for each of the twenty-one methods. These examples could no doubt be improved upon, and I should be delighted to receive suggestions for apter illustrations. Some of the difficulties of exemplification derive from the fact that many of the 'methods' refer to early stages of research and are consequently not readily apparent in 'finished' books and papers. In any case, the difference between the calmness of a text like this

TABLE 31. A classification of Moles's twenty-one[a] heuristic methods of scientific creation and infralogical modes of thinking

Ethics of thought	Heuristic method	Infralogical mode of thought
Traditional philosophy, 'as if'	Application of theory (1)[b] Transfer (7) Theory renovation (10) Recodification (14)	Analogical
Philosophy of 'no'	Hypothesis revision (3) Contradiction (8) Critical (9)	'Anti-logical' or logic of opposition
Concept of causes	Differentiation (5) Details (11) Phenomenal reduction (16) Emergence (19)	Logic of prolongation Mythopoetic logic
Philosophy of 'why not'	Combining two theories (2) Experimental disorder (12) Matrix of discovery (13) Dogmatic (17) Aesthetic (20)	Logic of continuity Logic of juxtaposition Verbal logic
Philosophy bound by conventions	Limits (4) Definition (6) Presentation (15) Classification (18) Emergence (19)	Formal logics (binary, probabilistic, polyvalent)

[a] By definition, the twenty-first method, that of general theorems, defies classification.
[b] The numbers in brackets refer to the 'method number' and order of discussion in the text.

Source: Based on Moles (1957, 179), with minor modifications, some due to the process of translation into English.

book and the thrill of the early stages of geographical research into a problem is quite profound, as Moles himself has declared: 'The difference . . . between published science (achieved science which one finds in the literature) and *scientific creation* is *profound* and *irreducible*' (Moles, 1957, 51; translated from the French).

The first ten methods are largely heuristic:

1. *Method of the application of a theory*. This method consists in applying an extant theory to a new area of the subject.

Example from geography: The Davisian 'geographical cycle' of erosion (Davis, 1899) was extended to periglacial regions by Peltier (1950) in the context of a climatic geomorphology (see further comment on climatic geomorphology in examples to methods 3 and 9 below). Of course, cycles are now *passé* in geomorphology, and in a recent eighteen-man survey of periglacial geomorphology, Peltier's 1950 pioneering paper receives but one fleeting mention (M. J. Clark, 1988, 64). Such are the cycles of research. (For 'a late twentieth-century perspective on the Cycle' see Tinkler, 1985, 172.)

2. *Method of combining two theories*. The originality of this method is greater to the extent that the two theories appear incompatible at first sight, but, if they are incompatible to a marked degree, there is a matching danger that the method will founder because of the different logical chains in question.

Example from geography: The Lowry model (I. S. Lowry, 1964) combined economic base theory and the concept of population potential to model 'the generation and spatial allocation of urban activities and land uses' (Hepple, 1981*b*, 196). Economic base theory divides urban and regional activities into basic and non-basic sectors, the former working for export. Various relationships between the sectors can be studied and extended. Population potential was developed from social physics' measurements of accessibility to points by given masses of people. As Hepple (p. 197) points out, in later developments of Lowry-type modelling, 'the allocation by population potential was replaced by gravity and entropy-maximizing formulations'; and a Lowry-type model could be based on input–output modelling rather than economic base concepts (Wilson, 1974, 243). Further combinations of theories (models) ensued.

Later stages of the Pittsburgh simulation [Lowry model] were concerned with extensions of the classes of land use to include commercial and manufacturing as well as retail and manufacturing employment and population . . . often called the *Pittsburgh urban renewal simulation*. The major contribution of the Pittsburgh urban renewal simulation to applications of simulation to urban analysis was the successful design and implementation of a simulation *that linked several submodels*, such as housing, land use, economy, and population in order to imitate overall urban development. (Catanese, 1972, 240; my italic; a convenient summary of the original Lowry model and its widely applied successor, the Garin–Lowry model, is to be found in Reif, 1973, 170–217; and in Wilson, 1974, 220–43.)

3. *Method of hypothesis revision*. This is oriented towards the deepening of existing knowledge. It is not a critical reaction to existing theories, but rather their consolidation by reference to other supporting principles.

Example from geography: Climato-genetic geomorphology is a development from climatic geomorphology (Büdel, 1982, Fig. 5, 14; 252 ff.).

4. *Method of limits*. This consists in the exploration of confused domains lying between two aspects of phenomena or two contrary concepts. *Natura non fecit saltus* (nature does not make jumps—though it does so under the microscope). Man invented measurement (continuity) and a move from an initial dichotomy to a more continuum-like concept is the dissolution of the qualitative into the quantitative.

Examples from geography: *Re* incompatible theories. The original climax in vegetation theory was a monoclimax. 'In polyclimax theory, all climax types are of equal rank rather than subordinate to the climatic climax as is required by monoclimax theory' (Matthews, 1985a). 'Others, notably R. H. Whittaker [1953] consider that . . . [climax vegetation] is made up of a continuum of plant populations which intergrade in response to complex environmental gradients' (Matthews, 1985b, 85). Still other ecologists believe that the stability implied by the climax concept is never achieved because of changes in environmental conditions. *Re* dichotomies. The urban centre/rural periphery boundary was recognized as the rural–urban fringe (T. L. Smith, 1937), as a zone (Blizzard and Anderson, 1952), as not lying between social dichotomies (Wirth, 1938), as a continuum (Miner, 1952), and as not being a continuum (Pahl, 1968). (For a tabular summary of the 'rise and transmutation of [views] of the urban centre/rural periphery boundary' see Bird, 1977*a*, 103–51.)

5. *Method of differentiation.* Investigation here concentrates on a shadowy area where two concepts are not easily distinguished. For example, the opposite of a high-fidelity transmission is not an absence of transmission but a less faithful transmission. Here we are in the realm of tolerance limits and acceptable thresholds, and consequently the method is very common in applied psychology.

Example from geography: Tolerance or non-tolerance of industralization schemes by locally affected communities seems a 'shadowy area' until it is realized that a

Group tolerance limit [can be] made effective by: (i) influence on political decision-makers and/or (ii) influence on business decision-makers (creation of a perceived hostile investment climate). (Text extracted from a diagram on 'The concept of tolerance limits to development', Clout *et al.*, 1985, Fig. 6.3, 88)

But the consequent lowering of the tolerance surface cannot be considered permanent. Permanence can only come through legislative modifications (which will spread the new lower surface through the national space) or through the introduction of new local regulations or plans. (ibid. 88)

6. *Method of definition.* This method cuts up reality with complete arbitrariness: the giving of names, the talk of 'structures'.

. . . the schematic nature of a concept is rarely an obstacle to its value. Later science will overtake and eliminate it when it has played its part . . . It hardly matters that the strong idea [*une idée force*, which gave rise to the definition] transforms itself into a false idea [*une idée fausse*], if in the course of the transformation it plays a constructive role. (Moles, 1957, 66)

The method of definitions appears to be like a vaccination against rambling thoughts. (Moles, 1964, 51; both quotations translated from the French)

Example from geography: Ogden (1967) in a linguistic and psychological analysis of the concept of 'opposition' used 'Town and Country' as an example of opposition by definition:

Town and Country. If this is taken as a typical case of opposition by definition (based on statistical density of population, houses, etc.), the value of the opposition in practical application is relative to the growth of suburbs. In due course the distinction might vanish altogether throughout the entire surface of an urbanized planet. An opposition originally created by definition (in response to factual requirements on the basis of a cut) is thus shifting to a scale whose extremes are being gradually obliterated by the expansion of its middle (suburban) range. A temporary stage is thus reached where semantic complications are produced by legal definitions in terms of difference rather than opposition. Finally, in such cases, the oppositional definition may retain historical significance only. (p. 78)

This purely linguistic and deductive approach to changes in the definition of town and country is matched by what geographers have discovered empirically: 'Where does the city end? Certainly the municipal boundary has little significance as denoting any sort of closure, except for administration and statistical space, which thus gives a false image because of its non-correspondence with the real world' (Smailes, 1971, 12).

7. *Method of transfer*. This is a most fruitful method and at the same time most dangerous and consists not only of transfer of concepts and theories to areas where they have not been applied before, but also, and more generally, all kinds of analogical procedures.

Examples from geography: Transfer of a concept. Couclelis (1986), in seeking a 'theoretical framework for alternative models of spatial decision and behavior [,uses] a framework derived from [Zeigler's (1976)] discrete model theory that clarifies the relationships between three seemingly different types of models of spatial choice (based on the principles of stimulus-response, rational choice, and cognitive information processing)' (p. 95).

Transfer by analogy. This happens every time a geographer employs a simile or metaphor, and is implicit in any use of general systems theory. The idea of the gravity model is an obvious borrowing from physics, as is the idea of centrifugal and centripetal forces in urban geography (Colby, 1933; Bird, 1977a, 105–7).

8. *Method of contradiction*. This could be called an exacerbated case of the constructive imagination (see Table 28), where one advocates the opposite of an existing concept, theory, or method. It is rare that outright contradiction appears in print, but, to the extent that the method leads to the formulation of an alternative, it is constructive. 'Because of the rules of the . . . game, it is rare that a contradictory point of view manifests itself in the resultant publication . . . which claims to "limit itself to the facts of the case" '

(Moles, 1957, 69). The method is evident in the attempted constructive criticism of unfavourable reviews.

Examples from geography: Coleman's (1985) *Utopia on trial* had a thesis which claimed 'that modernist housing architecture does not bring the benefits claimed for it, and is not even neutral, but exerts an actively detrimental influence which often results in problem estates' (Coleman, 1987, 115). Dickens's (1986) critical comments on this thesis included a contrary research strategy suggestion: 'To pursue explanation involves conceptualising "social malaise" and talking to those participating in it' (p. 298). The subsequent debate between Coleman (1987) and Dickens (1987) reveals two different views on the causes of deviant social behaviour; and study of this now polarized debate will no doubt help future work in this area. While many journals print critical and contradictory comments on papers, the *Geographical Journal* is the only major geographical journal to record discussion contributions from the audience made immediately after the delivery of a paper. When these contributions contradict the paper upon which they are commenting, we have not only the author's original statement in the paper and the contradiction but also the author's instant reply in the light of the contradictory criticism proffered. Such polarized debates are great stimulators of creative thoughts in the minds of readers.

9. *Critical method.* This differs from the previous method only in the degree of expressed opposition to extant work.

Example from geography: Goudie (1985, 84) gives eight 'limitations' of climatic geomorphology.

10. *Method of theory renovation.* Sometimes long-established theories need 'restoration', or a 'face-lift', because of new factors that have emerged in work carried out since their original promulgation.

Example from geography: The Turner frontier thesis (Turner, 1894) proposed

that an ongoing frontier process was the catalyst for the American experience and, as such, this peripherally situated phenomenon was the unique factor in the evolution of a distinctly American-style culture. (Block, 1980, 31)

Although Turner's theories now seem antiquated in light of modern analysis, his relationship to the development of American geography remains significant. (Block, 1980, 41; for 'Some of the more important analyses of Turner's work by geographers' see Block, 1980, 32 n. 3)

The following eleven (plus) methods are more adventurous then the first ten in that they are not variations on a pre-existing theme, but strike out in new directions. Sometimes they even look like a case of creation *ex nihilo*, though since memory is involved in creation, all the methods involve rifling the past.

11. *Method of details*. Here comes into play an 'applied curiosity' about lacunae in existing explanations and also a readiness to respond to the unmotivated nature of what may turn up in research.

Examples from geography: Example of a 'plugged gap'. A lacuna was recognized by Haggett (1965, 141–2):

Here we come to the essence of the unsolved set of problems that we may characterize as the 'Morris–Oxford', 'Ford–Detroit', or 'Carnegie–Pittsburgh' problem, i.e. the problem of why one rather than a set of apparently similar cities proved a successful launching pad for a great industrial venture. The answer probably lies outside the field of human geography and in the study of individualism and industrial opportunism . . .

This was written before the rise of behavioural geography in general, and before the behavioural study of industrial spatial decision-making in particular, as exemplified by McNee (1974), whose editor commented as follows: 'It is symptomatic of the neglect of this field, and hence of the novelty of much of this book, that as much as one-quarter of the terms used by McNee in Chapter 2 and suggested by him for indexing, do not appear in David Smith's recent and comprehensive book, *Industrial Location* (Wiley, 1972), (Hamilton, 1974, 13).

Example of the capricious manner in which details leading to ideas are 'thrown' in one's way. One of many instances of a happy chance occurred to me when I was undertaking research for a book on centrality (Bird, 1977a). I read Moles and Rohmer's *Psychologie de l'espace* (1972), and was so struck by the liveliness of the ideas that I decided to investigate what else these authors had written. This led me to Moles's *La création scientifique* (1957), a now long-admired work, and, of course, the source of this section on methods, itself an example of serendipity.

12. *Method of experimental disorder*. A gamester's mentality solicits chance by 'stirring up one's mind'.

Examples from geography: It is not often one can expect to find an author 'playing games' in finished work. Bunge (1966*a*, 216–22, Figs. 8.2–8.7) did so in order to demonstrate 'basic geometric concepts available to geography' (p. 216), and as a result the outline of the US went through instructive transformation: 'the transformations were gradually less constrained through several geometries. As we allowed . . . more and more "violence" in their behaviour we were stripping down to more and more basic geometries' (p. 222).

A computer game simulation to reconstruct accidental drift voyaging in the Pacific tested whether population diffusion in Polynesia was wholly planned or not and was one of the first such uses of the computer in which geographers took part (Levison *et al.*, 1969).

13. *Matrix of discovery*. We ask here 'What are the empty cells in an information matrix?' Even an arbitrary and simple, two-dimensional matrix can be very useful. '. . . the power of a concept, that is to say the seductive nature of its appeal to the intellect, has infinitely more heuristic value than its exactitude' (Moles, 1957, 80).

Example from geography: Berry (1964) proposed a regional synthesis via a geographic matrix. Clark *et al.* (1974) spotted a gap—'the flows between places, with one matrix for each flow category in each time period' (Johnston, 1979, republished 1987, Figs. 3.2–3.4, 79–80).

14. *Method of recodification*. If different words are used to describe concepts and phenomena, or perhaps to coin new collective nouns, by-products may result; or we may transform the qualitative into the quantitative by some form of mathematics.

Examples from geography: The word 'gatekeeper' is not merely a collective noun for estate agents, solicitors, and employees of banks and other financial institutions with resources to issue loans, but also indicates their functions as regulators in the allocation of scarce resources. It also suggests a possibility of change via 'micro- and meso-mechanisms' rather than the necessity of overturning whole social structures (Eyles, 1979).

Qualitative description of accessibility and networks can be rendered into more precise numerical indices by graph theory

(Garrison, 1960; Kansky, 1963; Tinkler, 1977), though network analysis has not provided the once-hoped-for synthesis between physical and human geography. And even the following, from a preface to a book on network analysis in geography, now looks optimistic: 'we argue that there are equivalent spatial structures common to both fields and that we can use common mathematical models for both kinds' (Haggett and Chorley, 1969, p. v).

15. *Presentation method*. Use of tables, graphs, schemata (explanatory, functional), and all kinds of diagrams (even maps!) fall under this heading.

Example from geography: Kauth and Thomas (1976) liken the development of crops in spectral data space, as seen by Landsat, to an *n*-dimensional envelope, rather like a woolly hat with tassels (see also Crist and Kauth, 1986, who 'de-mystify' the 'tasseled cap'). The cap represents a three-dimensional envelope, and the image of the tassels emerging from it represent possible further dimensions:

The Tasseled Cap concept simply involves identifying the existing data structures for a particular sensor and application (i.e. set of scene classes), changing the viewing perspective (i.e. rotating the axes) such that those data structures can be viewed most directly, and defining feature directions (new *x*-, *y*-, and *z*-axes in the cube example) which correspond to spectral variation primarily or exclusively associated with a particular physical scene class characteristic. (Crist and Kauth, 1986, 83).

16. *Phenomenal reduction* (called 'Méthode de réduction phénoménologique' in Moles, 1957, 99, because of links with Husserlian procedures). An artifical variation of the field of perception is brought into focus via detachment of phenomena from time contexts and from 'real' space (isotropic plain), and via new viewpoints (aerial and satellite surveys).

Examples from geography: 'Counterfactuals' in history and hence in historical geography compare what might have happened with what actually took place (Prince, 1971, 52–3; Norton, 1984, 55–7). A vivid spatial counterfactual is given by Rose (1966), in discussing the 'metropolitan primacy versus the rank–size rule' debate with reference to urban hierarchies. Here is Rose introducing his fictitious continent as a contrast to the actual Australia and its actual history of urban development:

Imagine an isolated continent, *à la* von Thünen, uninhabited, and unapproachable by surface to potential settlers because of the unnavigability of the surrounding seas. Call this land Continent A. . . .
Continent A is circular in shape with a diameter of 2,000 miles. Transport economics concerned with the exploitation and assembling of materials for export from the new continent make it rational to concentrate inter-continental rocketry facilities at one point on its surface and bring all materials to that point. In this matter, surface transport is far cheaper to use than rocket transport for all possible distances on the continent. Consequently, the initial point of contact is made with the new continent at the central location that comes to be known as Rocketport (RP . . .). For the agricultural settlers of Continent A, RP exists as the only significant market point on the surface of the entire continent. (pp. 6–7)

Rose's conclusion is that for such a continent, as for the states of the actual Australia, metropolitan primacy is the normal state.

17. *Dogmatic method*. 'Every logical construction is true, and to give each one practical value, one only has to find the domain of application' (Moles, 1957, 104; translated from the French). This is the logic of 'Why not?'

Example from geography: The shortest network of lines connecting a given set of points separated in space is the 'soap-bubble' pattern, and shows that the intersections within the connecting network may not include any of the original points (Miehle, 1958; Bunge, 1966*a*, 187–8). Therefore, if a costly (per route-kilometre) network is constructed (with a tendency to route-minimization of the 'soap-bubble' extreme case) between already existing nodes (i.e. already existing origins and destinations), then intersections will arise in areas separated from the original nodes. This 'logical construction' is borne out by formerly isolated junctions of the nineteenth-century railway network (sometimes giving rise to railway towns) and, in this century, green field motorway junctions, often prime targets for 'out-of-town' developers.

18. *Classification method*. 'The importance of classification is not intrinsic, either in the presentation of the table of facts classified (always to some extent false), or in its exhaustiveness as a procedure which is never complete. Classification is a creative stage, a sort of "pincer-grip", allowing apprehension of concepts in their evidential isolation' (Moles, 1957, 107; translated from the French). Sciences which make classification their entire aim

are destined to disappear as disciplines become more sophisticated (e.g. anatomy; botany, as primarily a taxonomic exercise; zoology, see Popper, 1970, 54, quoted in Bird, 1975, 161), but the construction of classifications and their development is often a fine learning exercise in newer areas of disciplines.

Examples from geography: The entirely classificatory geographies of the nineteenth century (called pejoratively 'capes and bays' geography) remain only as examples of extinct procedures, but classification still opens up new areas of understanding (see following).

A classification of glaciers was proposed by Andrews (1972) derived from a W_E/W_T ratio, where W_T is total glacial power and W_E effective glacial power, a proportion of the average glacial velocity resulting from basal sliding. The W_E/W_T ratio is small for polar and subpolar glaciers but larger for temperate glaciers.

19. *Emergence method.* Here we arbitrarily make discontinuities emerge from the continuities of the actual phenomena studied.

Example from geography: In the real world, away from coasts and other 'sharp' physical features, the boundaries of urban hinterlands are generally not linear, and there are considerable overlapping areas of urban fields emanating from different centres, depending on the indicators used. Arbitrary limits of cut-off can be proposed to produce linear boundaries; or, in the case where several indicators are used, we can propose a median boundary line as a synthesis (Green, 1955, using seven measures of the hinterlands of New York city and Boston); for elaboration of the term 'hinterland' in a port context see Bird, 1971, 124–47).

20. *Aesthetic methods.* Moles (1957, 117–18) was here thinking of cosmogony, very general theories, and appeals to aesthetic appreciation which emerges from subconscious realms. We also feel it to be pleasing if a simple concept is highly productive. As human geographers we study what Johnston has described as a 'pre-interpreted world of meanings' (Johnston, 1986a, 450); and some of the interpretations, leading to actions that have great spatial effects, may have been based on aesthetic criteria.

Example from methods used by geographers: It is aesthetically appealing to use a logical construction from a more basic discipline (physics, mathematics) as a foundation for a geographical theory. The gravity model of distance-decay, early transmuted into Reilly's law of retail gravitation (Reilly, 1931), has spawned a whole family of gravity-related models (P.J. Taylor, 1975), with a long run in human geography, not without attracting criticism (e.g. Jensen-Butler, 1972; Fotheringhame, 1981, 1982).

Examples of aesthetic methods used to produce phenomena studied by geographers: The site and master designs chosen for cities planned in this century often derive from aesthetic criteria, e.g. Chandigarh, India, (Khosla, 1971); Brasilia, where the original basic design was 'the drawing of two axes crossing each other at right angles, in the sign of the Cross' (Costa, architect, quoted in translation by Evenson, 1973, 143 and 146); and Canberra, where the Prime Minister of Australia in 1907 summed up the final factor in the detailed selection of the actual site: 'I do not say that picturesqueness alone should decide the question; but, other things being equal, I think that the beautiful ought to turn the scale' (Fitzhardinge, 1954, 11).

21. *Method of general theorems*. Such general theorems—attempts at synthesis of long-term value—are not very numerous, even in the whole of the sciences, natural, social, and human. Where they exist, they often provide a kind of intellectual crossroads.

Examples from geography (with help from outside the discipline): An obvious example is systems theory in geography. Other possible candidates are, in physical geography, process-oriented studies and energetics (K.J. Gregory, 1987); and, in human geography, structurally oriented studies, including Marxist geography and the abstract research of transcendental realism (see Fig. 13).

Methods 22 +. There are obviously many more different tricks of the trade for getting and developing ideas, or more than one way of skinning a cat (CAT standing for concepts and/or theories). The list of 'methods' above is, of course, not exhaustive. Moreover, we have reckoned without subconscious processes (Moles, 1957, includes a study of 'infralogical methods', see Table 31 above). As for choice of the more logical and conscious methods cited it really is a case of horses for courses; and, as on any racecourse, the recipe for success is judgement born of experience.

The real world (see also App. I.II, p. 247)

Surely, the real world is what geographers wish to study, along-side those images and schemata of the world which motivate significant actions in space. One sees how easily the 'real world' is overlaid with apriorisms, frames of reference, images, orientations, schemata, and *Weltanschauungen*, so that it seems ever to recede. How easy to slip into 'the idea of the real world'. No wonder that discussion of the subject so far has been nervously compromised by placing the 'real world' in defensive single quotes. Big issues of philosophical and psychological debate are involved, not excluding the fundamental idealism–materialism polarity. We must at the outset confront what I have called the 'idealist trap', and then briefly look at the statement 'orientation precedes observation' before proceeding to see how this impinges on studies of perception. Finally, 'perception of the real world' may be a subject whose study contains a lesson for the practice of geography.

The idealism trap is sprung by the assertion that one cannot prove that the material world exists wholly independently of minds. As far as I know, this is impossible to refute by using empirical evidence; but Popper has made a valiant attempt to defend objective reality: 'While no evidence can be conclusive we seem to be inclined to accept something (whose existence has been conjectured) as actually existing if its existence is corroborated; for example, by the discovery of effects that we would expect to find if it did exist' (Popper and Eccles, 1977, 10). The American pragmatist philosopher, William James, put it this way: 'The only objective criterion of reality is coerciveness, in the long run over thought. Objective facts . . . are real only because they coerce sensation. Any interest which should be coercive on the same massive scale would be *eodem jure* real' (1920, 67; originally 1878).

If we now proceed to the idea that orientations precede observations, and that those observations may well be coloured by the orientations, we must remember that this does not preclude the existence of an objective reality. Harrison and Livingstone (1980, Figs. 1 and 2, 25, 27) hold that presuppositions, 'the apriorisms implicit in all scientific investigations' (p. 26), arranged in a hierarchy, precede orientations, and influence every stage of the problem cycle (p. 29). If we transfer this sequence into scientific methodology, we again stress the primacy of deduction over induction, of 'making before matching' (a term coined by Gombrich, 1960, see his index). In a thought experiment Popper (in Popper and Eccles, 1977, 428–9) corrected the inductive beliefs of his co-author. Eccles proposed a thought experiment in which an observer was suddenly transported to the moon:

The external world, or the moon world in this case, is to him secondary to the way he comes to knowledge about it from his primary experiences which are delivered by his sense organs.

[Popper:] I do not agree . . . I do not, of course, want to deny that the senses are immensely important, and, as you have mentioned, this will be particularly true if an adult is suddenly placed in completely new surroundings. But even here I would wish to claim that we would first make a hypothesis as to where we were, and then try to test that hypothesis. In other words, we would use a trial and error, a making-and-matching process: a process of conjecture and refutation. This is why I think that the old story that the senses are primary in learning is wrong. . . . The senses have two roles: first they challenge us to *make* our hypotheses; second they help us to *match* our hypotheses—by assisting in the process of refutation, or selection.

Turning to perception studies, we find plenty of examples of those who believe that seeing is a deductive activity, via indirect realism, and the following are a selection of quotations from those who have held this view:

For it is certain maxim no man sees what things are, that knows not what they ought to be. (Jonathan Richardson, artist, 1715, quoted in Gombrich, 1960, 10)

. . . what he sees conveys no information until he knows beforehand the kind of thing he is expected to see. (Medawar, 1967, 133)

Perceptions as hypotheses. (R. L. Gregory, 1974, *passim*)

. . . we can see only what we know to look for . . . (Neisser, 1976, 20)

[A quoted] classic experiment shows the important role that prior experience or expectancy has on what we receive or recall. (Bourne *et al.*, 1979, 340)

. . . what anyone sees is not independent of his mental and verbal categories but is in fact a product of them. (Fish, 1979, 245)

The eye is not a passive but an active instrument, serving a mind that must be selective if it is not to be swamped by a flood of indigestible messages. (Gombrich, 1979a, 199)

Without a pre-existent framework or 'filing system' we could not experience the world, let alone survive in it. (Gombrich, 1979*b*, 1)

The most extreme form of indirect realism in perception studies has been called mental constructivism by Lombardo (1987, 318), and he cites Gregory and Neisser (see above) as modern exemplars of this position in a tradition stretching back to Berkeley and Kant.

It would be neat to rest upon the concordance between the hypothetico-deductive method of science and the deductive approach to perception. But there is the interesting case of a psychologist for whom perception studies were a life's work: J.J. Gibson (1904–80); and he offers a critique of a completely deductive emphasis in perception of the real world. From 1930–60 he subscribed to a psychophysical theory, attempting to free psychology from the dualism of mind and matter, viewing perception as a function of the environment. Later, this particular holistic approach was transformed into an ecological theory, owing much to systems thinking, stressing the reciprocity of perceiver and environment (Gibson, 1979; Lombardo, 1987, who chose this phrase as title for his book-length study of Gibson). A key concept in Gibson's change of orientation is the idea of 'affordances'. These

are relational properties of the environment and exist as *opportunities*, whether or not an animal wants to use them. . . .
One is reminded of the subject–object distinction, but the animal–environment distinction is necessarily reciprocal due to the dynamic features of the relationship. If the concept of affordances moves towards bridging mind–matter dualism it does so by connecting ecological structure (space) and ecological processes (time), or 'meaningful matter' with ways of life of animals. (Lombardo, 1987, 307–8; for a further discussion of Gibson's affordances see Reed, 1987)

As Costall has suggested (personal communication; see also Costall, 1981), perhaps there is some interlocking cyclic relationship between perceiver and the perceived real world not unlike that between predator and prey, with hypothesis before theory-laden observation playing a role within a total understanding of perception.

Reverting to geography and the real-world problem, we might apply a lesson from perception studies. In the real world are affordances or opportunities which we, as geographers, recognize as significant for our purposes, or which we recognize in the perceptions of those who make spatial decisions (the 'double hermeneutic'). The recognition of a valid opportunity obviously has positive advantages, but one must simultaneously recognize the conceptual limits predicated in that very act of recognition. This is because we have made a choice between what are regarded as 'opportunities' as distinct from alternatives not so categorized.

Popper's three worlds

The philosopher Karl Popper has distinguished three 'worlds' called World 1, World 2, and World 3. World 1 encompasses all the phenomena in objective reality 'out there'; World 2 is the subjective world of individual consciousness, inside our heads; and World 3 is the world of objective knowledge enshrined in concepts coded in some form of language (see Figs. 21, 22, and Table 32; Popper, 1972, 74; 1974, 1050–3). The 'idealism trap' looms again, and it could easily be argued that if World 1 is part of a philosophical system, it is itself a World 3 concept. Welch (1978, 26) mentions Popper's World 1 only to cite Tuan's argument (1974b, 215) that the 'question of objective reality is tantalising but unanswerable, and it may be meaningless'. Tuan had previously stated that even geometrical space is a sophisticated human construct. Welch concurs: 'space has no operational meaning without the existence of thought processes' (1978, 26). But we have already seen how Popper proposes that objective reality is a conjecture or hypothesis that tends to be confirmed by the evidence of our senses; and as a result it is offered as a firm component of his system as World 1. He also argues against a materialism which would assert that World 1 is closed to World 2: 'that physical processes . . . must be explained and understood . . . entirely in terms of physical theories' (Popper and Eccles, 1977, 51 ff.). Having argued the existence of the real world (World 1), we must beware of going on to assert that World 1 develops completely independently of World 2. This is to indulge in the 'psychological fallacy', as clearly pointed out by Dewey (1898):

The psychological fallacy (occurs when) we conceive the environment [World 1], which is really the outcome of the process of development, which has gone on developing along with the organism [i.e. World 2], as if it was something which had been there from the start, and the whole

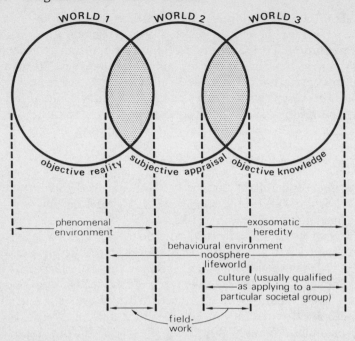

FIG. 21.   *Popper's concept of the three worlds*

Note: Exosomatic heredity refers to all the accumulated knowledge gained in the past and available to us and future generations, in contrast to bodily genetic, or *endosomatic,* heredity from our parents and their forebears; phenomenal and behavioural environments are terms discussed by Kirk (1963). For *noosphere* see Bird (1963; and index to this book). 'Lifeworld' derives from the *lebenswelt* concept of Husserlian phenomenology as interpreted for geographers by Buttimer (1976): 'Once aware of lifeworld in personal experience, an individual should then aim to grasp the shared world horizons of other peoples and society as a whole' (p. 281).

Source: Bird (1975, 159).

problem has been for the organism to accommodate itself to that set of given surroundings. (quoted, and the psychological fallacy further discussed, in Costall, 1986).

We may also note that in Popper's scheme of things World 3 interacts with World 1 only via individual World 2s, and this explains the relationship between the three circles in Fig. 21.

These matters have been discussed by Golledge (1979; relying on the work of Bergmann, 1957, and Berger and Pullberg, 1965; see also Moss, 1977, Fig. 5, 33):

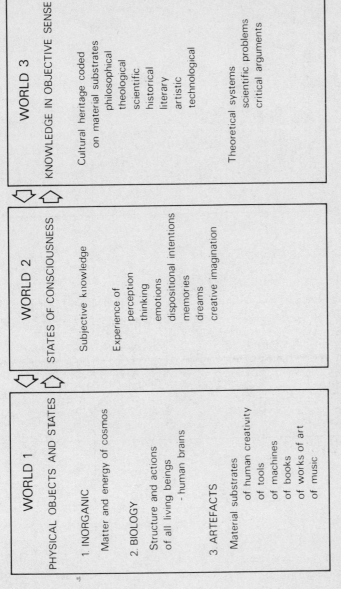

FIG. 22.   *Eccles's interpretation of Popper's three-world system*
Source: Popper and Eccles (1977, 359).

TABLE 32.   Popper's three-world system and Park's system of two different views of time

Worlds		Times	
World 1	Physical world, objective	Time 1	The time of physical theory, what is represented in the equations of dynamics by the letter t, what is registered by a clock, *tendency towards a determinacy polarity*
World 2	Subjective world of conscious experiences	Time 2	The time of human consciousness, process viewed from the standpoint of the present, *tendency towards a freewill polarity*
World 3[a] World 3.1	Objective knowledge The logical contents of books, libraries, computer memories, etc.[b]		
World 3 >> World 3.1	'... that is, world 3, transcends essentially its own encoded section'[c]		

Note: It is indeed a highly fortunate coincidence that Park appears to have independently chosen numbers for his two 'times' that are comparable with Popper's Worlds 1 and 2.

[a] '... [not] all the denizens of my (man-made) world 3 [are] man-made: the theories usually are, and so are the world 3 mistakes, but not the world 3 problems: they are internally generated, and perhaps discovered' (Popper, 1974, 1066).

[b] Books, libraries, and computers are World 1 objects.

[c] Popper (1974, ii. 1050; see also 1974, i. 143 ff.; 1976a, 180 ff.).

Source: for Popper, see nn. a and c; for Park, see Park (1980, 99 ff., esp. 100).

Realities are formed through an interchange of what each person thinks is inside or outside his/her organism and according to what the community thinks is inside or outside an organism. There is therefore *a dialectical relationship between individual and environment*, in which the environment affects and gives meaning to the individual who, at the same time, acts with respect and gives meaning to environment . . . This . . . is the process of objectification . . . (p. 116)

Golledge goes on to divide this process into four sub-processes, and it is possible to equate these with Popper's worlds (see the italics inserted in the quotation below). Golledge's four sub-processes of objectification are:

(a) a consensus of objectifications is the basis for institutionalization and the acceptance of a common external reality;
    [*an equivalence to World 3*]
(b) externalization separates self from external objects and allows external reality to have a structure independent of self;
    [*an equivalence to World 1*]
(c) internalization represents conscious awareness of an external object and the assimilation of the objectified experience into a belief system;
    [*an equivalence to World 2*] and
(d) historization is transmission of knowledge about objectifications among people;
    [*an equivalence to World 3*]
    (Golledge, 1979, 116; italicized 'equivalences' added).

The three-worlds system does not synthesize idealism and materialism, but allows each its domain to the benefit of the other.

The geographical literature has provided an example of how the three-worlds idea could contribute to the reification/scalar reductionism debate. Duncan (1980) attacked the 'superorganic mode of explanation in geography [which] reifies the notion of culture, assigning it ontological status and causative power. . . . In this theory culture is viewed as an entity above man, not reducible to the actions of individuals, mysteriously responding to laws of its own' (p. 181, from the abstract). Richardson (1981), commenting on Duncan's argument, pointed out that, in the eagerness to avoid reifying culture, we must avoid reducing it to simple interaction because the 'sin of reductionism is as mortal as that of reification' (p. 285). So the debate about culture proves a good example of the dangers of the extremes of reification and scalar reductionism. In Fig. 21 it is observed that culture is suggested as an example of a concept existing in World 3 of objective knowledge. While carefully avoiding the reification of culture, Richardson nevertheless advocates considering it as a concept outside individual consciousnesses, but able to impinge on them:

Humans interact through a medium of symbols. Central to such a mode of communication is objectivation [cf. the use of the term 'objectification' in the Golledge quotation above], which is the placement of our individual subjectivities into *matter located outside of our interior states* . . . symbolic communication forms *a holistic pattern located 'out there'* it is social intersubjective reality . . . like this text. . . . culture is a kind of 'talk', whose features include . . . a distancing that separates us from our 'talk'. . . . In addition, and of particular importance, *this mode of communication is located 'out there'*. (Richardson, 1981, 284 and 285; my italic)

The italicized words in this quotation seem to be entirely consistent with the idea of culture as part of objective knowledge in World 3, separate from the Worlds 2 of individual consciousnesses but able to interact with them. Popper's three-world system appears to help steer us between the extremes of reification and scalar reductionism as far as the important example of culture is concerned.

We shall also find that two of Popper's three worlds can help to clarify the kinds of change that occur in the discipline of geography. This 'never-standing-still' quality of any worthwhile academic subject is something to be finally confronted. But let it be admitted in advance of the exercise that any proffered solution should itself be subject to change and only on probation.

9

Change and Progress in Geography

I once knew a senior medical scientist who was so scathingly critical of everyone else's research that one wondered if his constitution precluded any possibility of belief. Considering his intelligence, he was expectedly barren of ideas of his own (something that may help explain his critical temperament), but when he *did* have an idea of his own—my goodness, was it not the most important and profoundly illuminating notion that the world had ever known. On this topic all his critical faculties were suspended: he was a complete sucker for his own idea; any opposition to it aroused a degree of resentment that fell not far short of active enmity.

(Medawar, 1979, 45)

Six types of change of geography in two worlds

There are two journals called *Progress in Physical Geography* and *Progress in Human Geography*, and a debate might rage over whether or not a discipline can progress in the sense of 'become better'. I follow Rescher's (1977) argument in rejecting any idea of 'real truth' to which we draw ever closer, viewing

the progressiveness of science as residing in the first instance not in the enhanced truthfulness of its later claims, but in the constantly augmented data-base on which its later theses take their probative stand: . . . (p. 175 n.) . . . this sort of self-appraisal is clearly not enough to underwrite adequacy in the face of an externalized view of the realities of the matter. Apart from how it fares from its own *internal* standpoint, there remains the more 'objective', duly externalized issue of the rational qualifications of the inquiry procedure. Room must be made for the operation of the *external* controls of a factor that is essentially disjoint from the realm of pure theory, viz., pragmatic efficacy.

We thus envisage a two-fold criteriology of cognitive adequacy (and *ergo* of scientific progress), namely the internal factor of systematization and the external factor of pragmatic efficacy. (pp. 176–7)

Such argument will need justification later in this chapter, but meanwhile there might be readier agreement with the idea that if an academic subject undergoes no change, it is moribund, if not dead. What are these changes? Here are some possible answers:

1. Change in our individual conceptions of geography;
2. Change in consensual conceptions of geography (the consensus applying to groups of various sizes located in physical proximity, e.g. 'national schools', or bound by some shared interest, or compartmentalized by a shared language);
3. Change in the real world;
4. Change in the real world as perceived by individuals and groups.

Some important topics lie embedded in the above list: the place of individual consciousness in academic study, the individual *vis-à-vis* the group, the acceptance of the real world as actually 'out there', and individual and group perceptions of this real world.

Popper's three-world system helps us to sort out these matters. We each have our own World 2 (individual consciousness), but access to other Worlds 2 is only via World 3 (results of knowledge published in some manner so that our own World 2 may assimilate it). Finally, there is World 1, which I now take unequivocally to be the real physical world actually existing out there (see previous chapter). Obviously, changes are occurring constantly in other Worlds 2, but we know of these changes only when their results appear in World 3. So if we are to suggest a tabulation of the kinds of change we may encounter in geography, we are concerned only with Worlds 1 and 3. The resulting six types of change in Table 33 require some amplification, which will include possible examples.

1. *Changes in World 1—endogenous.* Geography is one of those disciplines where the subject-matter itself changes (unlike a 'dead' language like Latin), i.e. change in the real world of World 1. This type of change refers to 'vertical' developments in World 1. These are the changes that take place *in situ* at the micro- (local) or meso- (regional) scales. They may be physical changes at the earth's surface following the laws of cause and effect—in short, the very domain of physical geography. Such physical changes may have impacts on the human geography of an area as hazards. The arrival of a new artifact within an area such as a dam may provoke

TABLE 33.   Six types of change in World 1 and World 3 of geography

1. Changes in World 1—endogenous: changes at the micro- to meso-
   (regional) scales
2. Changes in World 1—exogenous: resulting from World 1 influences
   flowing from one region (or regions) to another
3. Changes in World 3—so-called 'solutions' or predictions (always on
   probation) provoke attempted better[a] solutions
   and predictions, and/or new problems
4. Changes in World 3—new ideas often inspired by older ideas
5. Translation of ideas published in World 3[b] across space in World 1
6. Translation of ideas within World 3[b] to and from anywhere in entire
   academic curriculum into and out of geography

[a] 'Better' if they explain or interpret more of enlarged empirical data than previous
theories.

[b] Ideas obviously arise in the Worlds 2 of individuals, but they are not generally
available until published by whatever means in World 3.

change. But such examples as these also have behavioural dimensions, involving World 3: the hazard as perceived, and the decision-making involved in the construction of the dam; and the
debate about the latter may have been conducted outside the
area. Life is rarely endogenously simple, especially for the human
geographer.

2. *Changes in World 1—exogenous.* The area of World 1 being
studied may operate as if it were a closed system, perhaps in a state
of quasi-equilibrium, and then an influence from outside will
provoke internal changes. Floods are often caused by events
beyond the confines of the actual impact. The settling of lowland
England by the exotic Romans and Saxons paralleled the earlier
physical repercussions of the Alpine orogeny. Any system that
'cascades' across space, or, more generally, any phenomena that
undergo a spatial diffusion process fall into this category. The
boundary between this second type of change and the first type
alters with the scale of the study, i.e. diffusion from *outside* a
particular site may well be intraregional when study focuses upon
a wider area.

The remaining four types of change in Table 33 derive from our
developing thoughts about World 1 which are published in World
3.

3. *Changes in World 3—better solutions or predictions*. So-called solutions to problems or predictions about them provoke better solutions or predictions which explain more of an enlarged, empirical data base, part of what has been called the 'constant-revision' view of scientific progress (Rescher, 1978, 51). In behavioural geography the device of rationally optimizing man has been generally succeeded by satisficing actors with bounded rationality. A more detailed example for this category is offered by Jennings's rather dramatic research experience in confronting the problem of the origin of the English Norfolk Broads. Compare the following two quotations, the second published only one year after the first:

There is . . . good reason to explain the origin of the Broads in general, solely in terms of estuarine depositions and its consequences . . . (Jennings, 1952, 53)

Important new data have come to light calling for a major reconsideration of the earlier hypothesis of the origin of the Broads. Convincing stratigraphical evidence has now been obtained to suggest that all of the Yare Valley Broads are artificial. (Jennings and Lambert, 1953, 91; see also Lambert and Jennings, 1960)

Jennings was quick enough, and big enough, to see that new evidence had overturned his original solution, an example of the probationary nature of all scientific statements in the Peircian-Popperian version of the scientific method. Both these philosophers, seventy-six years apart, warn us to be on the alert to recognize when our ideas are wrong or outmoded:

. . . any scientific proposition whatever is always liable to be refuted and dropped at short notice. A hypothesis . . . is something which looks as if it might be true and were true, and which is capable of verification or refutation by comparison with facts. The best hypothesis, in the sense of the one most recommending itself to the inquirer, is the one which can be the most readily refuted if it is false. This far outweighs the trifling merit of being likely. For after all, what is a *likely* hypothesis? It is one which falls in with our preconceived ideas. But these may be wrong. Their errors are just what the scientific man is out gunning for more particularly. (Peirce, 1931–5, i. 48; originally *c.*1896)

. . . it happens more often than not that natural selection eliminates a mistaken hypothesis or expectation by eliminating those organisms which hold it, or believe in it. So we can say that the critical or rational method consists in letting our hypotheses die in our stead; it is a case of exosomatic evolution (Popper, 1972, 248; for 'exosomatic' see Fig. 21).

4. *Changes in World 3—new ideas.* New ideas are often brought into being by the desire to improve on extant ideas, or by comparing ideas. As we have seen, many of the twenty-one types of scientific creation expounded by Moles (1957) involve a form of comparison. Ideas when published often provoke their own supersedure, because rooted at their date of origin they invite comparison with what has happened since in World 1 and World 3. Ironically, the 'best' ideas, because most stimulating, are in danger of the quickest obsolescence due to the subsequent published thinking or rethinking they inspire. In hazard research the idea of narrow, technological solutions to problems of flood responses was replaced by a conceptual approach involving 'a range of possible adjustments' (White, 1961). This concept has been characterized as a 'dominant view', and has itself been subject to challenge as being too circumscribed (Hewitt, 1983, 12 n.).

Consensual recognition of societal structures (World 3) and also of the importance of the individual (World 2) may lead on to considerations of the transactions between them (Walmsley and Lewis, 1984, 45) in some form of the dialectic or reflexive modes of explanation. At the most general methodological level the recognition that geographers have employed contrasting modes of explanation provokes an idea of occupying the centre ground via the duality of structuration (Johnston, 1983a; Storper, 1985).

The last two types of change in Table 33 involve some form of 'translation', including transfer across space in World 1 (type 5 below) or of transfer into geography of ideas from elsewhere in World 3 (type 6).

5. *Translation of ideas published in World 3 across space in World 1.* Geography, like the phenomena, localities, and regions it studies, is an open system, open to the idea of the diffusion of solutions suggested for place X to be translated across space to place Y (translation across World 1). In the Wooldridge and Linton (1955) monograph on the shaping of southern England, itself now superseded, we can see a famous example of spatial translation whereby Linton's deductions about the Wessex rivers (see Fig. 11 above) were translated eastwards into the Weald and London Basin. Ullman (1967) called such a translation 'geographical prediction': 'This means that if one has no trend to predict in a particular area, one looks for essentially analogous conditions elsewhere, transfers the setting to the study area,

making appropriate adjustment [i.e. "translates"], and uses the results for prediction' (p. 131). The concept of the CBD (central business district) and its delimitation began in the US in 1954 (Murphy and Vance, 1954), and it took five years for the method to be applied to Australia (Scott, 1959).

6. *Translation of ideas within World 3*. This final type of change covers translation of ideas in and out of geography to and from anywhere in the entire university of the arts and sciences (translation into and out of geography within World 3). Agnew and Duncan (1981) have actually proposed a schema for evaluating such borrowings. One has only to think of physical geography's debt to physical sciences, human geography's to economics, and what geography owes to statistics to realize that illustrations of this type of change need not be laboured by detailed illustration. Nor is the traffic just one way; for example, geography is probably a net exporter of ideas to archaeology.

Physical and human geography and the problem of their integration

Johnston (1986a, 449–50) has pointed out that the word 'geography' has two usages, vernacular and academic. In the first usage, physical geography, for example, is the physical environment and is always with us; in the second usage, reference is made to academic objectives and practices, which is the meaning discussed in this section. Something strange has happened over the last twenty-five years. Physical geography can be studied without much reference to human activity, in a pure scientific manner, or in reference to those areas where man has made but little detailed impression: high latitudes and high altitudes, and in the very atmosphere above us. Yet we find that physical geographers have been more active than human geographers in attempting to bridge gaps that have opened up in the discipline. In their efforts to study relevant problems, many physical geographers have been forced to consider socio-economic issues; and while always using a variant of scientific methodology, some now avoid many of the features of positivism that have come in for the most trenchant criticism mentioned in Chapter 3; and these physical geographers realize that the 'book of nature . . . is to some extent a mirror: what looks out depends on who looks in' (Rescher, 1984, 185). Using systems approaches they have demonstrated the importance of human activity (e.g. K.J. Gregory and Walling, 1987). There have also been national calls for a holist approach to environment, symbolized by the US National Environmental Policy Act, 1969, and the European Communities Council Directive (27 June 1985, L 175/40), which, like the US Act, requires environmental-impact assessment for certain types of project from June 1988. This seems like a statutory requirement for an integrated geographical approach. Williams (1987, 388) noticed that in a discussion on the unity of geography, it was the physical geographers (Douglas, 1986; Goudie, 1986) who were more enthusiastic when focusing 'on the big questions of ecology' (Graham, 1986, 465).

The integrated study of 'territorial complexes' seems easier in a centrally planned socialist, socio-economic system, where, according to Anuchin (1977), such a view is not hampered by a private-ownership mode of production, and, in Russia, integrated regions are believed to have objective reality—the integration of their physical and human geography is taken as given: 'Soviet geography . . . clearly sees the objective character of the territorial complexes it studies. . . . Regionalization in Soviet geography is not only a method of inquiry but also a method of reorganization' (p. 142).

In the Western world, human geographers tend to cry 'Exception!' when the unity of physical and human geography is up for discussion. Physical geography and human geography

have different epistemologies, for one (human geography) but not the other is dealing with a pre-interpreted world of meanings; the subject matter is a human creation that is not subject to the operation of general laws. . . . I would not claim that the two vernacular geographies should be separated, for there is great merit in their symbiosis. But I would stress that they are different forms of science, and therefore not integrable. (Johnston, 1986a, 450)

'Different forms of science' is, I take it, a reference to different epistemological stances in human geography and a belief that the hypothetico-deductive procedures of science cannot be applied to much of the social science aspects of geography. Johnston has consistently argued this separatist theme, earlier finding no evidence for integration via the study of resources (1983c). In 1979b (p. 2) he had defended his use of the 'terms "geography" and "human geography" interchangeably'. This was because links between physical and human geography were tenuous; because the sharing of research techniques and research procedures was an insufficient foundation for a unified discipline; because of his own incompetence to write about physical geography; and because much of the human geography he was dealing with was North American in origin, and 'geographers there, especially in the United States, encounter no physical geography'.

Stoddart (1987, 330) attacked this 'sleight of hand [whereby] physical geography disappears altogether'. He also went on to attack those floundering Jeremiahs of geography who see geography invaded by each new social-science 'paradigm', overwhelmed by the assaults of yet one more -ism, so that eventually it undergoes de-definition (Eliot Hurst, 1980). What does Stoddart offer in place of such pessimistic forecasts for the future of an integrated geography? We might be able to agree with him when he asserts that the subject must be problem-oriented: 'The task is to identify geographical problems, issues of man and environment within regions' (p. 331). He then quotes the case of the geography of Bangladesh: '. . . there is no such thing as a physical geography of Bangladesh divorced from its human geography, and even more so the other way round' (pp. 331–3). But when we follow Stoddart further, we find the following:

My vision . . . is of a *real* geography—a reasserted *unified geography*, . . . and at the same time a *committed geography*. . . . It is a geography which reaches out to the future, and the future is even conditional on how well we do it. It is a geography which will teach us the realities of the world in which we live, how we can live better on it and with each other. It is a geography which we will teach our neighbours and students and our children how to understand and respect our diverse terrestrial inheritance. (p. 333)

Observe that geography is real, unified, committed, but we are not told exactly what it is, though for Stoddart such a geography obviously exists. It reaches out to the future, but are there changes as it does so? And although it teaches us the 'realities of the world', whose realities? If we have a diverse terrestrial inheritance, perhaps a geography with diverse approaches might be a better strategy, for then there might be opportunities for judging which approach is appropriate in which cases, especially if there is some external datum against which to test our endeavours.

If Johnston and Stoddart are taken as spokesmen for opposing views, we can acknowledge that they argue their cases cogently; but I hope to have shown that both their arguments have deficiencies, partially because both ignore the different levels of study in any one discipline. Up to first-degree level we can provide a common theme—'study based on curiosity concerning the world about us'—and we can introduce our students to a variety of perspectives from the natural sciences, the social sciences, and the historical human sciences (Fløistad, 1982). If we ourselves are expert in and advocate one particular perspective and one particular methodology, we must also expound the criticisms that our approach has attracted. Beyond first-degree level, like every other subject, research is problem-oriented, and many of the most interesting problems are in the overlap area between two or more disciplines (as arbitrarily defined for the undergraduate and school curricula, though undergraduates should get a whiff of this 'problem-ferment'). Another 'level' is represented by 'What are we talking about and at what level of generality?' Three of my physical geography colleagues have commented on the perennial problem of the relationship between physical and human geography in these very terms:

If we have the courage, we might suggest that the relationship between human and physical geography will not lie primarily at the level of either content (since the objects of study occupy a common world, but share relatively few attributes of behaviour) or technique (which, despite some real overlaps, is dominantly subject-specific) and thus must be sought in the realms of methodology and philosophy. (Clark, Gregory, and Gurnell, 1987, 383)

We shall see later in this chapter whether it is possible to sustain an argument that a link does exist between physical and human geography at the methodological-epistemological level.

Desiderata and dangers in attempting to define geography

In 1973 I speculated on the desiderata that might be necessary in attempting a definition of geography (Bird, 1973). Attempting to list the desiderata for a definition turned out to be more thought-provoking than the definition itself, which was a derivative language-assembly job. Today, as in 1973, not many care for subject definitions: they smack of fiats, party lines, Words of God. As soon as a fence is put round a subject, everyone inside will at least look out beyond, and the adventurous are certain to find a way through. Up to first-degree level there are geography curricula, and if the results of our research work figure (or will figure) nowhere in any of these curricula, not even as a footnote to a footnote, we must wonder whether or not we are 'doing geography'. Having said that, we must concede that in research we are problem led, and may heed the valuable advice of J.L. Piveteau (transmitted by Claval, 1967), who recommended a Janus (or double) outlook: 'One is . . . led to the necessity of a double definition: on the one hand, a description of things as they are and which delimits an acquired area of knowledge; on the other hand, an expression of the strivings, the plans, and the aims of the researcher' (p. 38). This is the idea of an academic subject as a continuum emanating from the past and going on into the future, 'a sort of geography-without-end-Amen' (Burrill, 1968, 11). Thus geography is what geographers have done; geography is what geographers strive to accomplish. In any case an overt time perspective is necessary to satisfy both historical geographers and geographer-planners. In surveying what geographers have done, it is ever-instructive to study past definitions, to see if they have anything of value for the present time. Already, four desiderata for a possible definition of the subject have so far surfaced: open-endedness, a double viewpoint, time perspective, and eclecticism. So heady is the delight in desiderata erection that another could be added to act as a brake, in the form of 'conciseness', but not at the expense of clarity. The definition ought to be in explicit terms.

If geography is problem led, perhaps we could borrow Gombrich's phrase and say that the methodology is some form of 'making and matching' (Gombrich, 1960, see his index): venturing a guess or a hypothesis about the problem, having a hunch, and then seeing if this has any utility by some form of test against the data. Such statements will not please everyone, but they indicate a broad class of approaches under the heading 'hypothetico-deductive method'. In 1973 I rather skated over the 'scientific method', which, as we have seen, can cover a whole family of approaches, but it might be useful here to flesh out the basic Popperian stages of the hypothetico-deductive method with comments assembled from the work of the American philosopher John Dewey (Table 34). We still retain the problem orientation, but we

TABLE 34.   The Dewey formulation of the 'theory of inquiry'
compared with the basic fivefold stages of Popper's hypothetico-
deductive method

Popperian stages	Dewey's theory of inquiry
[P_{-1} Problem orientation]	Antecedent conditions
	Situation is indeterminate; doubtful (and we are doubtful); confused (outcome cannot be anticipated); conflicting (evokes discordant responses)—a complex of factors united by a pervasive quality. 'The peculiar quality of what pervades the given materials, constituting them a situation, is not just uncertainty at large; it is a unique doubtfulness which makes the situation to be just and only the situation it is. It is this unique quality that not only evokes the particular inquiry engaged in but that exercises control over its special procedures.'[a] '. . . the immediate existence of quality, and of dominant and pervasive quality, is the background, the point of departure, and the regulative principle of all thinking.'[b]
P_1 Initial problem	Institution of a problem
	'The first result of evocation of inquiry is that the situation is . . . adjudged to be problematic.'[c] 'The questionable becomes an active questioning, a search . . . The scientific attitude may almost be defined as that which is capable of enjoying the doubtful; scientific method is, in one aspect, a technique for making productive use of doubt by converting it into operations of definite inquiry.'[d]
TT Tentative theory	Determination of a problem-solution
	Against hard and fast distinctions between observation and conceptual formulation, between fact and theory, between the perceptual and conceptual. '. . . perceptual and conceptual

TABLE 34. Continued

Popperian stages	Dewey's theory of inquiry
	materials are instituted in functional correlativity with each other, in such manner that the former locates and describes the problem while the latter represents a possible method of solution. Both are determinations in and by inquiry of the original problematic situation whose pervasive quality controls their institution and their contents.'[e]
EE Evaluative error elimination	Reasoning
	'An hypothesis, once suggested and entertained, is developed in relation to other conceptual structures until it receives a form in which it can instigate and direct an experiment that will disclose precisely those conditions which have the maximum possible force in determining whether the hypothesis should be accepted or rejected.'[f] 'In order to gain . . . [theoretical] understanding, we must not only envision new possibilities, but we must explore their systematic interconnections with the complex network of our concepts.'[g]
P_2 Remaining problem(s) different from P_1	Operational character of facts–meanings
	'Ideas are operational in that they instigate and direct further operations of observation; they are proposals and plans to bring new facts to light and to organize all the selected facts into a coherent whole.'[h]
	'. . . knowledge can . . . be characterized as the warrantably assertable product of inquiry. . . . When knowledge is achieved, the specific inquiry is completed, but in a new situation these end-products may serve as a means for further inquiry.'[i]

a Dewey (1938, 105).
b Dewey (1960, 198).
c Dewey (1938, 107).
d Dewey (1929, 228).
e Dewey (1938, 111).
f ibid. 122.
g Bernstein (1966, 108–9).
h Dewey (1938, 112–13).
i Bernstein (1966, 110).

Source: The table has been compiled from Bernstein (1966, 101–13), a chapter collating and commenting on Dewey's writings on the theory of inquiry. The Popperian sequence, $P_1 \rightarrow TT \rightarrow EE \rightarrow P_2$ can be found with discussion in Popper, 1972, 119: P_{-1} has been added.

might well ask: 'Problems about what?' Hartshorne (1959) has performed a valuable service in providing the precious phrase 'the earth as the world of man'. More care has gone into this than meets the eye, especially the selection of the preposition 'as', which included the fields of perception and behavioural geography long before they became fashionable:

It is also necessary to pin down the elusive preposition 'as'. We study the earth not as something related to man and hence only in the ways in which it is related to man, but as an object in itself. But the scope of that object and the selection of phenomena included in it are determined by our basic interest in the object as we experience it (p. 47).

Nobody could forget Hartshorne's conception of that race of literate insects who succeeded our wiped-out human race, who were able to read our books via minimizers, and who came to the conclusion that our geography had missed the distinctions about terrestrial phenomena that were critical for ants if not for men (1959, 45). 'Terrestrial phenomena' is a useful phrase, for it seems to imply a distance-decay of interest away from the surface of the earth, but allows study of sub surface rocks and the upper atmospheres, all considered as part of the world of man.

As we survey the ocean of methodological literature in geography, is there anything else that cries out for inclusion in a definition? Facing a sustainable charge of subjectivity, may I plead first for some recognition of both the formal (uniform) and functional (nodal or polarized) approaches, a bifocal view of regionalization (classification) that has proved so fruitful. Secondly, there are the problems posed by the great range of scales that a geographer may employ. There are probably other desiderata for a definition, but extending the list will run even further counter to a vital consideration already included—conciseness. It is now appropriate to list the twelve desiderata already discussed:

1. open-endedness
2. double viewpoint
3. time perspective
4. eclecticism
5. conciseness
6. explicit terminology
7. hypothetico-deductive method
8. relevance to man
9. relevant subject-matter
10. formal approach
11. functional approach
12. scale problems

This was the list drawn up in 1973 with the exception of number 7, which originally read 'deduction and induction with emphasis on the former'. I probably should have stopped at that point, but I went on to assemble a definition based on the above desiderata (see numbers in brackets): *Geography is the scientific study (1, 2, 7) of changing (3) spatial relationships (10, 11) of terrestrial phenomena (9) viewed as the world of man (8, 4)*. This statement meets desideratum number 5, but in doing so 'relationships' are understood to cover formal and functional spatial patterns in both physical and human geography so desideratum number 6 is lost, as is number 12. Fifteen years ago I 'rather inclined' to this definition, while admitting it was neither elegant nor punchy, and ruefully conceded that any definition of the subject lies imprisoned upon the printed page, while time and the subject move on.

One commentator on my definition, R. Clark (1974), pointed out that, when looking at the phenomena studied by geographers, one gets a centrifugal impression. This view is always with us as exemplified by this report: 'Dr Compton said that as regional geography has disappeared and the discipline grown more diverse it had been difficult to discern any core curriculum within undergraduate programmes. Academics more and more view geographers more properly concerned with the preserve of other disciplines' (McQuaid, 1987, 9). But Clark did find centripetal features in the attributes of phenomena that geographers consistently study, and he quoted Nystuen (1963, 373):

Geographers have a common subject matter which reveals itself in certain words used again and again. The problems found interesting and being investigated in all branches of the discipline are defined using this common set of words—the controversy over the definition of geography notwithstanding. These words describe spatial arrangements and associations of activities and processes in geographical space. We adopt a spatial point of view whether the problem considered is one of physiography, cultural diffusion, economic expansion, or any of the diverse problems found attractive to geographers. Some of the words I refer to are: distance, pattern, relative position, site, and accessibility. Many others come to mind. [Clark, 1974, 156 added 'boundary, density, coincidence, discreteness'.]

Despite all the difficulties and dangers, geographers will always go on trying to give definitions of their subject. If you find my 1973 suggestion too positivist, here are three others, respectively from the standpoints of humanistic geography, historical materialism, and transcendental realism:

Humanistic geography achieves an understanding of the human world by studying people's relations with nature, their geographical behaviour as well as their feelings and ideas in regard to space and place. (Tuan, 1976, 266)

Geographical knowledge deals with the description and analysis of the spatial distribution of those conditions (either naturally-occurring or humanly-created) that form the material basis of social life. It also tries to understand the relations between such conditions and the qualities of social life achieved under a given mode of production.

The above could stand as a definition, but the author (Harvey, 1983, 189) expands on 'geographical knowledge' in the following paragraph, thus:

The form and content of geographical knowledge depend on the social context. All societies, classes, and social groups possess a distinctive 'geographical lore', a working knowledge of their territory and of the spatial distribution of use values relevant to them. This 'lore', acquired through experience, is codified and socially transmitted as part of a conceptual apparatus with which individuals and groups cope with the world. It may be transmitted as a loosely-defined spatial–environmental imagery or as a formal body of knowledge—geography—in which all members of society or a privileged elite receive instruction. This knowledge can be used in the quest to dominate nature as well as other classes and peoples. It can also be used in the struggle to liberate peoples from so-called 'natural' disasters and from internal and external oppression.

The third definition following appears to be within the perspective of the transcendental realist, but I am not sure whether the author would so class himself: 'The uniqueness of regions is now to be understood as socially constructed, the result of a synthesis of place-specific characteristics with more general social processes: reciprocally how these more general processes unfold is influenced by regional specificities' (R. Hudson, 1987). But there is a possible reaction to all these would-be definers of geography, myself included. 'We can possibly see *how* you arrived at your advocated definition, but you don't tell us why you adopted the particular *Weltanschauung* that colours the steps leading to your formulation.' We have to take the discussion of what geography is beyond the strait-jacket of any definition. If that all sounds rather stiff, let us put the matter in the American vernacular: 'Say, definer-man, I mean—Where's the epistemological beef?'

Towards a geographical methodology-epistemology: introducing PAME

Here we reach the key question of metamethodology:

Quis custodiet ipsos custodes [Juvenal, vi. 347–8]—By what method are we to control the choice of our philosophical methods themselves? (Rescher, 1979*a*, 17)

This book has attempted a critical survey of many of the concepts and methods used in geography, trying to assess advantages and disadvantages of many proposals. But I must here repeat the essence of the criticism levelled at Johnston (1980) by Eyles and Lee (1982, 117): one epistemology can hardly make valid criticisms of another because each has different criteria of validation. In turn Johnston (1986*b*) used this argument to criticize the following question by Paterson, as a positivist making a positivistic assessment of an historical materialist approach: 'in what way [may] a Marxist urban geography . . . be more "successful" than more conventional positivist approaches in analysing and explaining urbanisation under advanced capitalism[?]' (Paterson, 1984, 176). Paterson replied to Johnston's criticism as follows:

Johnston neglected to put the word 'successful' in quotation marks. I was aware, in asking that question, that it could be construed as a positivist question, yet it seemed to me that it was necessary to raise the issue of comparability of the two perspectives. In the book, I go on to observe that comparative studies need to probe to the philosophical roots of research. However, there is an important sense in which philosophy and theory are rooted in our experience of everyday life and are 'tested' by that experience. (Paterson, 1986, 113–14)

Here are some step-by-step comments on this debate, as I summarize what has been held to be valid and invalid criticism:

1. If a critic with a particular epistemological stance adversely criticizes work using that *same* epistemology, he is attempting to expose logical flaws within the methodological-epistemological edifice.

2. If a critic with a particular epistemological stance adversely criticizes work based on a *different* epistemology, the cry is raised that the critic is using different criteria of validation compared with those used by his object of criticism. The criticism is declared invalid because epistemologies are incommensurable.

3. A particular epistemology may be improved by 'internal' criticism (1 above). It protects itself from attack by criticism based on a different epistemology (2 above). There is no room for criticism based on some *external* datum.

4. Without possibility of reference to criticism based on some external datum, an epistemology is likely to be a self-supporting perspective, perpetuating itself by its own inertia, and progressively diverging from the problem area it was originally designed to illuminate.

5. An epistemology should therefore contain suggestions for criticism of itself against some external datum. Paterson adumbrates a possible solution when he suggests that 'philosophy and theory are rooted in our experience of everyday life and are "tested" by that experience' (ibid.).

Step 2 above arose in Kuhn (1970, 150): that what scientists observe depends on their paradigmatic perspective. I believe that Kuhn did a disservice by focusing on the sociology of practitioners (followed in geography by P.J. Taylor, 1976; and by Johnston, 1979, 184–5, with his generational model), rather than concentrating on the developing logic of what they do (World 3) (Haines-Young and Petch, 1980, 73). Kuhn's view can lead to extreme claims of critical inviolability for a perspective, as we have seen in the case of humanistic geography (Pocock, 1983, 358); or at least there is the claim that paradigms (in the sense of *Weltanschauungen*, constellations of beliefs, epistemologies) cannot be compared (Mair, 1986, 348; following Kuhn, 1970, 109–10). This is not correct, as the following example shows.

The imbalance-in-humors theory of disease in ancient Greek medicine and the bacterial-infection theory of modern medicine are incompatible as theories and may even be said to be semantically incommensurable. But when they are judged as inquiry procedures for diagnosing an illness and prescribing a treatment they become pragmatically commensurable. (Thompson, 1978, 494)

This matter will have to be borne in mind as we now embark on a route to a suggested methodology-epistemology for geography. In outlining this route I have sought aid from philosophy, for writers on concepts and methods in any discipline must be conscious of already flowing streams of thought. I can be accused of merely finding signposts for where I wanted to go in the first place. 'The eye . . . sees only what it seeks' (Marsh, 1864, 15). In answer to such criticisms I can but remark that the route of reasoning here presented corresponds to my experience as a professional geographer so far; and there is an active role for criticism in the suggested methodology-epistemology.

We can at least take off from a position of 'almost-consensus'. A possibly large majority of geographers would currently agree that the discipline is problem-oriented. Problems arise from contemplation of what has been achieved so far, and doubts about that achievement. The way forward is by some comparative process between belief about a possible solution to the problem in question and what we can ascertain about the problem and its context. As soon as we start to go into details about the 'comparative process' we lose consensus. My own preference is for some form of the open-ended version of the scientific method. Before anyone cries 'Positivist!', let me assert that so far nothing has been said about epistemology. For the last thirty years, geographers who have been promoting various -isms have usually started from epistemological bases, and only then do they go on to a deduced methodology to serve those fundamental -isms. My early instincts were to agree with this, in conformity with my general preference to go from the general to the particular in deductive fashion.

But this *epistemology first, then derived methodology* route has been condemned by post-modernists, including Dear (1988); and this leads into a logical difficulty.

Postmodernism is basically a revolt against the rationality of modernism. It is, in the specific context of this paper, a deliberate attack on the modernist epistemology. This epistemology is 'foundational' in character; it searches for universal truth and meaning, usually through some kind of metadiscourse or metanarrative. Postmodernism . . . holds out for a philosophic culture freed from the search for ultimate foundations or the final justification. . . . It has attacked the intellectual conditions which allow for and tolerate the dominance of one discourse over another. It has, therefore, worked against the potentially repressive power of theoretical metalanguages which can act to marginalize a nonconforming discourse.

Postmodern philosophy has been powered by a simple but penetrating question: 'On what basis can a claim be made for a privileged status of one theoretical viewpoint over another?' The essence of the postmodern answer is that all such claims are ultimately *undecidable.* (pp. 265–6)

Post-modernism is therefore good at deconstructing modernism, revealing 'conscious and unconscious strategies of exclusion and repression, . . . [which] are rife with internal contradictions and suppressed paradoxes' (Dear, 1988, 266). The problem then arises as to how a discipline is to be 'reconstructed', if the merits of the claims of any theoretical viewpoint, or grounding epistemology, are undecidable. Yet Dear does go on to define human geography as the description and explanation of 'the structure of society over time and space' (p. 271), via the three primary processes of 'conflict, production and exchange, and human interaction which characterize every human society to some degree' (p. 270). This is all nice and clear, but the more powerfully one advocates the criteria for a reconstruction of human geography, the more modernist is the flavour of the argument. Dear himself recognizes this logical impasse.

By defining the 'primary processes' which structure time and space, I have, in effect, taken a step toward defining some criteria by which my position can be justified. *I believe this to be unavoidable* if we are to circumvent the absolute relativism implied in deconstruction. The definition of evaluative criteria is still some considerable distance from defining a fully-fledged grand theory (an option which I explicitly reject). I am simply admitting into the conversation a more limited relativism in order to expose the basis for my judgements. (p. 270; my italic)

This does not seem to be very satisfactory because we are not told the grounds upon which we are to accept Dear's 'definition of evaluative criteria'.

As we shall see, the logical objection to putting epistemology first has been recognized for a very long time. But there is a way forward. Having heeded the advice of the post-modernists and decided not to put epistemology first, we now embark on a new route to avoid the logical impasse encountered above. The following argument recommends putting methodology first.

A typical view today sees philosophy at the base of a pyramid. It is thought that the approach to an adequate geography involves, first, the solution of the most general ontological and epistemological problems. . . . The view that I am advocating turns matters over. . . . this view suggests that geographers ought first to get down to business, to do geography, and then along the way reflect on what is being done. What is being assumed about the nature of the relationship between mind and body? What is being assumed about the nature of causality and intention? These questions arise in reflections on concrete work in a way that can be more productive than the way they do if they are attacked before the beginning of that work. (Curry, 1986, 93–4)

This is precisely the argument of John Dewey, as the following commentary by Bernstein demonstrates:

It may be thought that Dewey has approached the 'problem of knowledge' from the wrong end. His approach does reverse one of the major ways in which the problem has been attacked by traditional philosophers. One might think that we must first discover what are the distinguishing characteristics of knowledge, and then ask how this knowledge is to be achieved. What distinguishes genuine knowledge from mere fancy is the way it is related to the methods for discovering and testing it. To speak of knowledge divorced from the context of inquiry is to rip 'knowledge' out of its natural context. A good deal of the sterility of traditional epistemology results from this false abstraction. There are no intrinsic features that serve to mark off genuine knowledge: it is in the context of inquiry that we find the criteria for evaluating knowledge claims. (1966, 110–11)

This is not to say that 'geography is what geographers do', which I have attacked as a 'semi-circular statement' (Bird, 1973, 201). Completing the circle reveals the inanity: 'Geographers do geography.' Instead, the Curry argument, as I understand it, is that a proven methodology leads to an epistemology. I believe this to be the case, not because of a hunch or unsupported predilection, but because in the path of the sequence labelled *epistemology first, then derived methodology* there lies a formidable logical objection: the diallelus, or Wheel Argument.

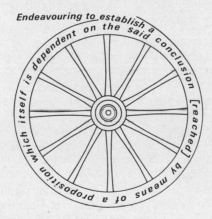

FIG. 23.   *The Wheel Argument or diallelus*

Top: The Wheel Argument (term coined by Rescher, 1977, 17) in a form of words based on a translation of an extract from an essay by Montaigne (1588, 1596, republished 1926, vol. 4, 111), the original of which runs as follows: 'Pour juger des apparences que nous recevons des subjets, il nous faudroit un instrument judicatoire; pour verifier cet instrument, il nous y faut de la demonstration; pour verifier la demonstration, un instrument: nous voila au rouet, (for translation see text).

Bottom: Based on the Oxford English Dictionary definition of 'diallelus'.

Rescher (1977, 15–18) has drawn attention to the problem of legitimation for cognitive methods presented by the Wheel Argument and presents this translation of its expression by Montaigne: 'To adjudicate [between the true and the false] among the appearances of things we need to have a distinguishing method; to validate this method we need to have a justifying argument; but to validate this justifying argument we need the very method at issue. And there we are, going round on the wheel' (see Fig. 23, top). In the face of the 'seemingly devastating *diallelus* argument' Rescher (1977, 23) advocates not scepticism 'but merely the less drastic conclusion that one should interpret the quality-control factor of the "success" of a cognitive method or criterion in the *practical* rather than the theoretical mode'. There is a nineteenth-century forerunner of this conclusion in a paper by the American philosopher Charles Sanders Peirce (1839–1914) entitled 'The fixation of belief' (1931–5, v. 223–47) in which he stated that inquiry is the struggle to overcome doubt and attain belief (pp. 231–2); and he compared four methods of inquiry, awarding the palm to the method of science. Modern philosophers have both praised Peirce's general endeavour and found flaws in his detailed reasoning, and it is worth quoting one commentator (Scheffler, 1974) at length:

Science does not establish, fix, and conserve theories, but exposes them continually to maximal test; the theoretical agreements of one period are accordingly often, in fact, uprooted and superseded by conflicting agreements at a later time. And this fact seems, in itself, incompatible with the project of showing science to be maximally effective in fixing belief, in general. Perhaps, it may be conjectured, some such train of thought accounts for the first difficulty with the essay [by Peirce] earlier mentioned, i.e., that the defence of the method of science seems to shift ground, moving from a consideration of effectiveness, to other reflections of various sorts.

These reflections seem to me to have this in common; they transfer attention from the stability of beliefs to that of methods, arguing that the method of science is *itself* firmer than the other methods discussed. Because it rests on the undoubted supposition of real things, because it is self-corrective, because it tests beliefs not by reference to human attitudes, intuitions, or institutions but rather by reference to those facts to which the beliefs in question purport to refer, scientific method is *itself* capable of standing firm through the change of specific beliefs. A challenge to a particular belief sanctioned by any of the other methods calls the method itself into question, because none of these methods is capable of allowing consistent correction of its own pronouncements. These methods are thus, one might say, brittle, incapable of absorbing changes without fracture. Or, changing the figure, they might be compared to a car without shock absorbers, in which shocks to the wheels are transmitted directly to the frame. The method of science, by contrast, achieves stability through flexibility. Rejecting pretensions to certainty, opening the testing process to the ideal community of all competent investigators, requiring continual correction to account for all available facts, the method is itself capable of absorbing change without upset.

This array of considerations indeed represents a shifting of ground in the essay [by Peirce], and it must be admitted that the essay does not therefore fulfil its promise. Nevertheless the defence of science it offers is of interest in its own right. Moreover, it represents an important emphasis on the *primacy of method* which is pervasive among pragmatic writers. (pp. 73–4).

Pragmatic writers! Have we been led to yet another -ism, that of pragmatism? This school of philosophy founded by Peirce has had some famous proponents, mainly American (including William James, 1842–1910, John Dewey, 1859–1952, G. H. Mead, 1863–1931, F. C. S. Schiller, 1864–1937, C. I. Lewis, 1883–1964, W. V. O. Quine, 1908–   , and N. Rescher, 1928–   ). Thayer (1967, 435) has attempted to sum up pragmatism's place in modern philosophy:

. . . it has helped shape the modern conception of philosophy as a way of investigating problems and clarifying communication rather than as a fixed system of ultimate answers and great truths. . . . The measure of success pragmatism has achieved in encouraging more successful philosophizing in our time is, by its own standards, its chief justification. To have disappeared as a special thesis by becoming infused in the normal and habitual practices of intelligent inquiry and conduct is surely the pragmatic value of pragmatism.

This summing-up is a reflection of the fate of epistemological pragmatism, or thesis pragmatism, which has run into several objections. Rescher, 1978, has summarized these as follows:

If it is *practical* utilities that are at issue, then the bridge from utility to truth is broken. There is no reason of principle why acceptance of a falsehood should not be enormously benefit-conducive, or why something that is benefit-conducive should be true. There may be *some* positive correlation between the truth of propositions and the practical utility of their espousal, but this is by no means enough to warrant taking the one as the criterion for the other.

He avoids such criticisms and the methodological scepticism of Feyerabend (1975) by advocating *methodological* pragmatism, a systems-theoretic approach to the theory of knowledge.

S.J. Smith (1984) complained that pragmatism has remained 'curiously neglected' by geographers (p. 353; but see Frazier, 1981) and pointed out that it has 'influenced Robert Park, an important figure in the history of urban social geography'. She also fought her way to realizing that pragmatism is important at the methodological level rather than primarily at the level of epistemology and ontology:

[Pragmatism focuses] not on the ontological level of what is believed, by faith, to exist, but on the practical level of what can be achieved by intelligent intervention in an imperfect and unequal world. . . . it is an action-oriented philosophy, stressing not only the inevitability but also the responsibility of analysts' intervention with a studied environment, and asserting the importance of formulating concepts in use, rather than in the abstract where they might have no practical relevance. (p. 369)

Though pragmatism has largely disappeared as a philosophical doctrine, as the above quotation from Thayer demonstrated, we can heed Rescher's (1977) call for a *methodological* pragmatism; and, in order to sell a form of pragmatism to geographers, it is necessary to go beyond reliance on pioneer figures in social geography and a hope that this particular -ism will provide us with a foundation epistemology. We have to trace out why a route to a pragmatic methodology will lead us to a useful linked methodology-epistemology.

To concentrate on methodology is to encounter such epistemological questions as 'What do you fundamentally believe?', 'What are your methodological presuppositions?', 'What is your *Weltanschauung*?' Rescher links methodological pragmatism to an epistemology via a double circle, in a figure-of-eight configuration (Fig. 24):

As this diagram clearly shows, the over-all justification is doubly self-sustaining. Two interlocked 'cycles' are involved: the 'truths' generated by the inquiry procedure must be sustained in implementing action, and they must themselves undergird the *Weltanschauung* by which the linkage of successful implementation and truthfulness can be validated. The

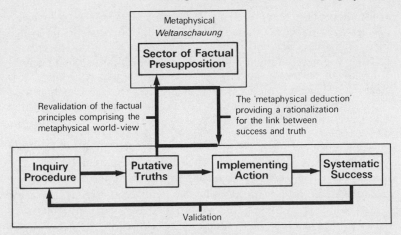

FIG. 24.   *Rescher's double circle of pragmatic justification*

Note: '. . . justification of an inquiry procedure lies in the closing of two cycles which constitute a double circle of the figure-eight pattern shown [outlined in heavier lines above] . . .' (Rescher, 1977, 107). For further explanation, see text.

Source: Redrawn from Fig. 4 in Rescher (1977, 107).

central requisite of the whole process of justification is that this complex interlocking dual-cycle should be properly closed and connected. The present theory of cognitive validation is literally a matter of 'wheels within wheels'. Only when everything is adjusted and readjusted so that *all* the pieces fit in a smooth dovetailing, will we obtain a workable pragmatic methodological justification for an inquiry procedure. (Rescher, 1977, 107)

In the course of a book-length study of competing *Weltanschauungen* in the social sciences, Wisdom (1987, vol. II) considers the problem of whether changes occur first in the *Weltanschauung* or in the discovery of new empirical theories at the level of methodology: 'What I suspect is that both procedures are possible, though, in nearly all historic examples I am familiar with, the new empirical theory has come first, and brought the new weltanschauung in its train' (pp. 80–1). This subordination of the epistemological level, suggested in the very title of Rescher's book from which Fig. 24 is drawn—*methodological* pragmatism—has been pushed further by C. Taylor in his paper 'Overcoming epistemology':

It becomes evident that even in our theoretical stance to the world we are agents. Even to find out about the world and formulate disinterested pictures, we have to come to grips with it, experiment, set ourselves to observe, control conditions. But in all this, which forms the indispensable basis of theory, we are engaged as agents coping with things. It is clear that we couldn't form disinterested representations any other way.

But once one takes this point, then the entire epistemological position is undermined. (1987, 476)

Rescher does not go as far as abolishing epistemology though his schema ties it in closely with methodology. He goes on to point out that we can indeed alter several elements in the schema: the *Weltanschauung*, the inquiry procedure, and consequently the putative truths of our hypotheses and we can also alter our implementing action. 'But the one thing that we cannot control are the *consequences* of our actions: . . . Here we come up against the ultimate, theory-external, thought-exogenously independent variable. Pragmatic success constitutes the finally decisive controlling factor' (Rescher, 1977, 108). One recalls the aphoristic poem of Ambrose Bierce:

> *A lacking factor*
> 'You acted unwisely', I cried, 'as you see
> By the outcome.' He calmly eyed me.
> 'When choosing the course of my action,' said he
> 'I had not the outcome to guide me.'
>
> (Bierce, 1963, 68)

An epistemology closely intermeshed with a methodology itself controlled by an external controlling factor avoids the blockage presented by the Wheel Argument. The external controlling factor is the correspondence theory of truth: empirical verifying fulfilment of expectations by correspondence to the theory in the real world; and the route to that correspondence is via the pragmatic success of the consequences of our actions based on our theories.

This pragmatic methodology-epistemology can now be considered in relation to geography. The methodology is some variant of the hypothetico-deductive procedure, such as outlined by Rescher (see Fig. 24), or by Popper compared with Dewey (Table 34). If we turn to Table 35, the methodology-epistemology is applied without difficulty to physical geography, and then we follow it downwards via the vertical line in the left-hand column. The question now arises as to the point at which we strike difficulties in

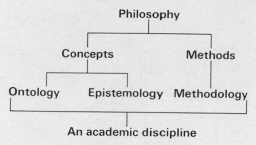

FIG. 25. *Suggested relationships between terms used in the philosophy of an academic discipline*

Notes: Throughout this rather arbitrary schema presuppositions are embedded, and a problem is: 'What is the justification for the presuppositions explicitly (or implicitly) held?' Suggestions for a definition of geography often remain merely at the concepts and methods level without justification of the underlying epistemology and associated presuppositions; and this criticism applies to the definition in Bird (1973).

For examples of ontology, epistemology, and methodology in geography see Table 9; and for another schema of the relationships between the terms see the 'presuppositional hierarchy' of Harrison and Livingstone (1980, Fig. 1, 27).

'Concepts' can obviously occur at different levels of generality (including within a discipline), but the position in this diagram is to stress that ontologies and epistemologies are indeed concepts and not immutable 'givens'.

the social science aspects of geography. Rescher (1977, 291–2) is careful to point out that his schema applies in the cognitive realm but not in the normative realm: in areas which are 'welfare-conducive', but not in those which are 'excellence-conducive', such as aesthetics, ethics, theology, and in areas of existence where there are 'enhanced desires for a life that is rewarding and meaningful'. In these areas we may take Bartlett's (1958, 197) suggestion that the criterion of pragmatic success is replaced by satisfaction with an attained standard. Because many sectors of human geography are not concerned with aesthetics, ethics, and higher spiritual aspirations, there is no bar in those areas to a pragmatically validated scientific methodology and its moderated and moderating epistemology. But since man does not live by material welfare alone, judgement of a different sort is necessary as to when to moderate the scientifically based methodology by a form of intersubjective validation. This point is reached when we arrive at the horizontal line in Table 35. Wisdom (1987) says that a scientific methodology ('metascience') can be used throughout all

TABLE 35.   A pragmatic analytical methodology-epistemology (PAME) for geography

Methodology-epistemology	Area of application	Test
A combined pragmatic methodology (based on the hypothetico-deductive system) and epistemology, mutually correcting each other into conformity (see Fig. 24).	Physical geography	Self-correction by successive scientific procedures and methods using a 'constantly augmented data-base'[a] and pragmatic success** of theories and methods.[b] (In technology predictive success is essential, often a matter of life and death; in science we may learn more from failure of predictions[c])
	→ Applied physical geography via algorithms derived from the methodology-epistemology	
	→ Repetitive features of human geography (in fields of economic and practical welfare)	

**

Can above methodology–epistemology be used?	Covering law controversy	Non-repetitive features of human geography (e.g. historical geography) and fields of human geography where non-material values and qualitatives norms important	Attainment** of a satisfactory (or satisficing)** standard[d]
--------→ ?			

** Indicate four points where subjective judgement is to be exercised, such as at what point does the solid vertical line become the pecked (‑‑‑?) vertical line. (Wisdom, 1987, would answer: 'Never', for he replaces 'methodology' throughout by 'metascience', vol. i, p. vii; this 'unity of method' in the natural and human sciences has been called 'naturalism', see Entrikin, 1985, and App. I.1, p. 247.)

[a] Rescher (1977, 175 n.): '. . . the superior epistemic standing of . . . [one] thesis [over another] does not reside (in the first instance) in the superiority of its content, but in the superiority of its probative method'.

[b] 'The issue of cognitive adequacy and the legitimation of the instrumentalities of knowledge-acquisition leaves wholly untouched the question of the inherent rationality of our uses of knowledge' (Rescher, 1977, 289).

[c] This greater necessity for confirmation seems to distinguish technology from science. 'Science is to understand, technology to do. But applied science, though a step on the way to do something, is itself an extension of understanding' (Wisdom, 1966, 370). Agassi (1966, 349) puts it aphoristically: 'The excessive demand for confirmation is the tool by which the complacent postpones the implementation of novelties.'

[d] The suggestion that in areas less amenable to a combined pragmatic analytical methodology-epistemology the test criterion is a qualitative consensus on the attainment or otherwise of a satisfactory standard is derived from Bartlett (1958, 197), see note to Table 28.

social sciences, and therefore would deny the necessity for that horizontal line and the hesitancy of the vertical line.

If we attempt to apply the included hypothetico-deductive method to all parts of human geography, we shall run into controversy. This can be illustrated by debate over the covering law model in history, and hence historical geography (Norton, 1984, 4–5). The model was first put forward by Popper (1934; and later by Hempel, 1942, as made clear in a footnote, Popper, 1962, ii. 364). In its simplest form it states: a combination of initial conditions (C) and an appropriate law (L) gives resultant events (E). Thus, for historical explanation of events, descriptive laws are required: the hypothetico-deductive method. Historians have resisted, preferring the 'model of the continuous series'. To understand something historically it is necessary to *trace the course of events by which it came about*', a 'continuous series of happenings' (Dray, 1957, 68); and Nowell-Smith (1981) also depicts historical explanation as narrative, but the narrative of a complex web of the conceptual and empirical and not to be confined by the canonical form of the covering law (pp. 147 ff.). As we have seen, in their turn, historical geographers have sought for 'deep processes' in historical materialism or some other form of structuralism. The controversy continues (see the discussion in Wisdom, 1987), but we owe a debt to Harvey (1969a), who in the course of his major references to covering laws (pp. 36, 48–54, 75–9) unearthed (p. 51) this pertinent comment by Joynt and Rescher (1961, 153):

. . . it is clear that the historian in effect reverses the means–end relationship between fact and theory that we find in science. For the historian *is* interested in generalisations and *does* concern himself with them. But he does so not because generalisations constitute the aim and objective of his discipline, but because they help him illuminate the particular facts with which he deals.

I have compromised by suggesting that the position of the horizontal line in Table 35 is one for judgement. Judgement has even to be employed for pragmatic success, as the following debate shows. Rescher's 'methodological pragmatism' as part of an integrated system (including epistemology) has been criticized on the grounds that pragmatic success cannot play a direct role in the cognitive system (Bonjour, 1979). Rescher's reply includes the following:

Success and failure in the pragmatic/affective realm do indeed make a characteristically 'direct' impact on our beliefs. The practical failures of life's 'school of hard knocks'—the pains, discomforts, and distresses that ensue when things go wrong are peculiarly difficult to decouple from their characteristic belief-impact.

To say all this is not, of course, to gainsay the plain truth that pragmatic success and failure do not feature *directly* in our belief-system, but only operate there via *judgments* of success or failure. It is simply to stress that the coupling between actual and judged success—and above all, between actual and judged *failure*—is so strong that no very serious objection can be supported by pressing hard upon a distinction that makes so little difference. (Rescher, 1979*b*, 174)

In Table 35, the 'area below the horizontal line' and the 'pecked vertical line' allow for alternatives to a pragmatic methodology based on some variant of the hypothetico-deductive system in human geography; and there is demonstrable methodological and epistemological pluralism in human geography, which also reflects back its critiques on natural-science modelling so prevalent in physical geography. This critique of scientific methods is healthy because nothing could be more anti-scientific than a totalitarian unitary methodology. As Popper has averred (1986), the greatest disadvantage of Darwinism as a theoretical system is that it has defeated every rival so far with the result that there is no compeer and adherents with whom Darwinists could engage in profitable debate.

Now I sum up the advocated pragmatic analytical methodology-epistemology (PAME) (see also Table 36) and make

TABLE 36. Ten-point summary of the route towards a suggested methodology-epistemology (PAME) for geography

1. The Wheel Argument (see Fig. 23) necessitates a search for validation in the practical (methodological) rather than the theoretical (epistemological) mode.

2. *Internal* validation of methodology by augmented data base on which claims of later theses are compared with earlier theses; *external* validation of methodology by correspondence to the theory in the real world—its pragmatic efficacy. This is how a discipline progresses.

3. 'Truths' (always on probation) generated by the inquiry procedure, validated above, must undergird epistemology, which in turn revalidates methodology. Adjustment and readjustment of methodology and epistemology to ensure smooth dovetailing.

TABLE 36.   Continued

4. Leads to a pragmatic (combined) methodology-epistemology (PME).

5. Applies to cognitive realm of welfare-conducive areas (material well-being) but not to normative realm of excellence-conducive areas (see next point).

6. In excellence-conducive areas (such as aesthetics, ethics, theology), pragmatic success is replaced by a consensus criterion of satisfaction (no doubt temporary) with an attained standard.

7. For geography add an 'analytical' concept to suggested pragmatic methodology-epistemology = PAME: to re-emphasize methodological pragmatism rather than thesis pragmatism; to avoid, in human geography, any danger of the reification of structures; and to suggest more emphasis on a *shared* human experience, but avoiding also thereby anti-theoretical concentration on individuals.

8. Result is pragmatic analytical methodology-epistemology (PAME), with the hyphen there to emphasize the interlocking, mutually adjusting features of the system.

9. In some areas of human geography where PAME applied with less facility, note the criticisms of other methodologies and epistemologies. (Some criticisms will emanate from perspectives where political conviction overrides the Wheel Argument and where 'truths' are not on probation.) Employ humanistic illustration as a detailed contrast to hypothetico-deductive procedures.

10. As indicated in references throughout this chapter, the above sequence 1–5 owes much to Rescher (1977), but I have added a suggestion from Bartlett (1958, 197; see Table 28) for point 6 above. The suggestion for use of the term 'analytical' (point 7) comes from Couclelis and Golledge (1983). Other routes to warranted knowledge, particularly in some areas of human geography, will place the structure of PAME into relief, either by outright adverse criticism, or by light thrown from an entirely different perspective, possibly assisting the continual adjustment and readjustment integral to PAME.

reference back to the generalized triad of positivism, humanism, and structuralism illustrated in Fig. 8 above. Each word of PAME and the included hyphen are necessary and in the indicated order; so let this summary proceed one element at a time:

PRAGMATIC. This word comes first to stress the importance of external validation by correspondence with the real world, comparing consequences with the outputs of our work. Had epistemology come first we should be whirled around on the Wheel Argument (Fig. 23). Mair (1986, 348) follows Kuhn (1970, 109–10 and 150) in alleging that there are no ' "rational" metaparadigmatic tests for adjudicating theories'. This is the Wheel Argument in disguise. As we have seen, this argument can be circumvented by pragmatic tests—experiencing the results of using methodologies, paradigms, theories.

ANALYTICAL. This is inserted to oppose any approach that *starts* epistemologically from a posited structure. In social theory it is not difficult to find disadvantages both in the reification of structures on the one hand and in scalar reductionism to individuals out of context on the other (Table 15 above). But the reification of structures seems to me the greater danger, not only because we may easily become pawns of our creation, but also because, by putting epistemological structures first, we are confronted by the Wheel Argument. In detailed reference to transcendental realism (Fig. 13 above), 'analytical' is taken to rule out the suggested 'abstract research', and in the suggested 'concrete research' of this form of realism, the hypothetico-*deductive* method would be applied to many or few cases in their structural context. This is because I am opposed to a methodology which uses consideration of individual cases *inductively*, in the Micawberish hope that something will turn up to confirm the over-arching structures already in place.

As regards the relationship of PAME to positivism, we can first agree with Hay (1979, 2) that there are many varieties of positivism, actual and alleged (see a summary of the many forms of positivism in Couclelis and Golledge, 1983, 332). For some authors, the word has become a blanket term of abuse to label advocacy of any form of scientific method in the social sciences. Thus PAME may be called positivist by polemicists who are not impressed by the argument. But 'analytical' includes no claim to 'autonomy' by use of the hypothetico-deductive method—that is to argue that the investigator is objectively detached from the data with a value-free ideology. As Mair (1986, 347) points out, Kuhn (1970, 150–1) was among the first to demolish the idea of the objective scientist. But we can take the lead offered by Couclelis and Golledge (1983) that positivism has metamorphosed into an analytic tradition which places more emphasis on the human aspects of our world, and on 'open, public, intersubjective tests of knowledge by *continuous reference to experience*' (p. 334; my italic). If this meaning placed on 'analytical' leans away from structuralism and towards humanistic geography, it is towards a *shared* human experience, avoiding the stress on individuals that is at the heart of some anti-scientific and anti-theoretical claims of many phenomenological and existentialist approaches (see Couclelis and Golledge, 1983, 338 n. 7).

Couclelis and Golledge (1983, 335) give 'three reasons of principle' for choosing the analytical mode of discourse:

First, analytic languages are the closest we have yet seen to a truly public interpersonal/intersubjective (if not entirely intercultural) mode of communication; . . . Second, it is possible for analytic discourse to be explicit about its own limits in a way no other mode of thinking is likely to be. . . . Third, because the gist of analytic discourse is purely formal, that is syntactical, no particular semantic commitment about the real is called for.

METHODOLOGY. Methodology precedes epistemology, and is some form of the hypothetico-deductive method, which can be closely monitored by comparing outputs with experience of consequences. This is why it is qualified by PRAGMATIC. These outputs are Kuhn's 'exemplars', a 'basic and important' meaning he ascribed to 'paradigm': 'concrete puzzle-solutions', 'employed as models or examples, can replace explicit rules as a basis for the solution of the remaining puzzles . . .' (Mair, 1986, 350, quoting, Kuhn, 1970, 175). Mair points out that geographers have neglected this meaning of 'paradigm' in favour of another, less important use by Kuhn in describing a pan-disciplinary matrix: 'entire constellation of beliefs, values, techniques' (Kuhn, 1970, 175).

THE HYPHEN IN METHODOLOGY-EPISTEMOLOGY. The hyphen is there to stress the interlocking, mutually adjusting features of methodology and epistemology in the Rescher system, reproduced as Fig. 24 above; and I repeat a part of Rescher's description: 'Only when everything is adjusted and readjusted so that *all* the pieces fit in a smooth dovetailing, will we obtain a workable pragmatic methodological justification for an inquiry procedure' (Rescher, 1977, 107). The one stable element is the hypothetico-deductive nature of the methods of inquiry. And because the method is pragmatically warranted by successful correspondence to the thing in the real world, all else in the methodological-epistemological structure can be changed as experience dictates.

A position resembling part of the PAME strategy was suggested for geography by Chouinard et al. (1984), based on the Lakatos (1978) idea of scientific research programmes. Where these coexist, as they do in geography, 'the criteria for accepting or rejecting a particular research tradition in human geography must be programmatic, that is they must evaluate the coherence and creativity of a particular approach in terms of expansion in the conceptual and empirical scope of explanation' (1984, 375). This is very reminiscent of Feyerabend's diagrammatic exposition of the Popperian version of the scientific method (see Fig. 1 above); but it refers only to an internal validation and rejects (see 1984, 348) '*external* validation of methodology by correspondence to the theory in the real world—its pragmatic efficacy'—a quotation from Table 36, point 2.

Any advocacy of an approach which includes a pragmatic element (where judgement of success is based on success of results) lays itself open to the command: 'Show us results.' It would be easy to demonstrate successes of PAME (using a hypothetico-deductive methodology) in physical geography, but I myself have used such an approach in an apparently 'humanistic' area—the interviewing of decision-makers (see Table 24 above; and Bird *et al.*, 1983). I took notice of my own suggestion from point 9 of Table 36: 'Employ humanistic illustration as a detailed contrast to hypothetico-deductive procedures.' This was done by providing extensive quotations from the interviews, giving prominence to this in the titles of two reports: 'Transport decision-makers speak' (Bird, 1982) and 'Freight forwarders speak' (Bird, 1988). The pragmatic test of the research lies in the successful completion and reporting of eighty and seventy-two interviews respectively in two research projects. While humanistic illustration was prominent, I was in no doubt that all the information was collected against hypothesis-directed enquiries (intrinsic to the questions asked—see Table 24 above). This led to the usual hypothetico-deductive 'no-lose' situation: if the hypotheses are confirmed by the interviews, happy is the researcher; if hypotheses are refuted by respondents, then even more interesting questions may arise as to why this should be so.

Though highly detailed humanistic illustration can enliven a hypothetico-deductive strategy by its very contrast (just as microscale approaches can illuminate macroscale approaches and vice versa), PAME cannot be eclectically amalgamated with approaches under the general banners of humanistic-geography or structuralism. Apart from contrast, other epistemologies will supply criticisms of PAME which, if justified, will accelerate the 'adjustment and readjustment' that is integral to its progress. Geographers must judge, as PAME competes alongside those other epistemologies (and methodologies) for attention and adherence. Only the unknowable future will disclose the relative fortunes of all these suggestions in the life history of geography (arbitrarily presented as a staged history in Table 38).

While a PAME-based geography may carry a commitment to a course of action, decisions to implement actions are taken in the political realm, and, presumably, they are 'externally' warranted by electoral probation. But this does not mean that geographers, physical or human, escape the political implications of what and how they study (for a few examples see Table 37).

Open-ended conclusion

Developments in space, developments in time, and developments in thought all enrich our subject, not because any change necessarily involves qualitative progress, but because such developments generate a wider agenda, with greater opportunities for internal as well as external comparisons. I remain unconvinced that all academic study is proceeding to some omega point of absolute truth and am rather pleased to understand that a majority of physicists still believe in the 'no model' theory of the 'Copenhagen Interpretation' of 1929 (see index entries to this in Zukav, 1980). This holds that there is no ultimate, absolute, true model of explanation underlying the physical universe.

Some may believe that a barrier to progress in geography as a unified discipline may derive from the fact that it employs methods in the natural, social, and human sciences, a width of approach shared perhaps only with medicine. I have long believed that this can be a source of strength through stimulating comparisons of both subject-matter within World 1 and methods within World 3. This point of view found some support in a survey of 'all types of sciences', 1966–78, and the following quotation repeats the very last words of the introduction:

TABLE 37.   Possible political implications of some terms (arranged as polarities)[a] used by geographers in the course of attempts to explain or understand spatial patterns and relationships

More 'left'	More 'right'
Social equality	Economic efficiency
Dispersal (i.e. subsidies for peripheries)	Centralization
Central planning	*Laissez-faire* policies
Subjectivity	Objectivity
Society	Individual
Socio-economic structure	Agency (individualistic)
Structural determinism	Voluntarism
Nurture	Nature
Short-term evolution (or revolution)	Long-term evolution
Anti natural-science models (even for natural science)	Natural-science modelling (even when studying phenomena of natural science)[b]

Note: The political polarities are expanded into three perspectives: 'free-market', 'welfarist', and 'marxist' by Jones and Moon (1987, 341 ff.) and applied to medical geography.

[a] Complete adherence to some of these polarities may be impractical in that the extent of the compromises (trade-offs) between polarities may well be part of the art or science of politics.

[b] This is because such activity may well support the political status quo; but this applies only in so far as the society is not 'left dominated'.

TABLE 38.   An arbitrarily dated three-stage life history for geography

Stage	Date	Characteristics	Principal questions
1 Pre-scientific	Pre-1966	Descriptive generalizations; stumbling block: phenomena (including locations) seen as unique	If scientific methods not possible, how can generalizations be hierarchically related to each other?
2 Scientific	1966[a]	Process—(cause and effect) oriented	Can scientific methods be successfully extended to all human geography?
3 Plurality of methodologies and epistemologies	1972[b]	Science continuing, especially for physical geography, alongside rejection of science (called positivism by critics, using that term pejoratively) in favour of other -isms	Can geography avoid disintegration?

Suggested continuation by this study

			What are the methodologies-epistemologies that guide advocated approaches?
4 Pragmatic analytical methodology-epistemology (PAME)	Present–?[c]	Number of -isms seen as healthy; democratic approach to methodology-epistemology, including importance of oppositions to mainstream, all subject to continual testing	
5 ?	Future	?	Is the proposed methodology-epistemology externally validated[d] and as a consequence capable of self-correction?[e]

[a] Date of Bunge (1966b), 'Locations are not unique'.

[b] Date of Harvey (1972a), his first Marxism-inspired paper.

[c] If this suggested fourth stage is shown, with evidence, to be erroneous, the question marks in stage 5 can be replaced with another view of the currently advocated methodologies and epistemologies, with supporting evidence.

[d] External validation can be conclusions of inquiry compared with consequent events, comparison of predictions with results (correspondence theory of truth), or of a study's judgemental standard compared with a later consensus standard (with consensus perhaps aided by hindsight).

[e] Such a question is entirely avoided by simply continuing to describe all geography as pluralistic and leaving it there, i.e. continuing in Stage 3 of this suggested 'life history'. 'Pluralism' appears to me as a weak assertion, merely offering a pseudo-objective description rather than an account of modern geography based on an argument with an overtly declared perspective that can be debated with and challenged.

Perhaps the historical human sciences may learn something from the methodological rigour of the natural and also the social sciences. And the natural, and perhaps also some of the social sciences, may be reminded that they too, are in the end human sciences at once contributing to but dependent upon the values and preferences at any given time inherent in human understanding. (Fløistad, 1982, 17)

It might be best to base any criterion of quality for an academic discipline not so much on the solutions it proffers in the teaching curriculum but rather upon the problems the discipline currently identifies for the research intellects of the future. We seem to have been building bigger mental gymnasia in which to exercise those intellects, with greater debate about the appropriateness of various epistemologies.

Society is placing more demands on geography. It is no use merely offering to identify problems in the physical and social environments, intellectually demanding as that might be. 'We also want solutions for our money.' It is fair to reply that in a postmodern world there are no panoramic panaceas, and the words 'final solution' will never lose their terrible echoes. But we practise a discipline which understands where the laws of cause and effect rule in the world about us and where the actions of our fellow human beings, by centuries of effort to lift themselves above the sheer state of survival, have caused areas of our lives—including areas of our very spatial existence—to rise above mere mechanical application of cause and effect; and these areas of our life may be among the most precious to us. We can offer a range of solutions based on the lessons that geography has already learnt, and be ready for the pragmatic test of the consequences of actions based on what we say. When we fall short, we must relearn and adapt. This emphasis on change in geography is paralleled by the changes in the society where it is set; and society influences the geographers, who are inescapably part of it. Table 39 presents a series of headlines describing three generations of people in western Europe. We can see similar developmental change in geography: 1950s, 1960s: 'the quest for spatial order'; 1970s, 1980s: 'revolutions everywhere, pluralism'; 1990s: 'make pluralism productive'.

This last headline for the 1990s might well find expansion in what Soja (1987, 290) calls the 'postmodernization of geography':

TABLE 39.   Headlines describing three generations of society in western Europe[a]

1950s	1970s	1990s
Sow	Reap	Plough
Sacrifice	Self-gratification	'No nonsense' (no miracles)
Repairing war damage	Individual rights and freedoms	Rights *and* duties
Common goals	'Me era'	Interdependence
What can I do for my country?	What can my country do for me?	What sort of country do I want to live in?[b]
Co-operation	Confrontation	Network society[c]
Enlarge the cake	Divide the cake	Enjoy the cake
Prudence	Over-confidence	Resilience

[a] Based on de Vries (1988, 3) and derived by him from the 'Hartog model'.

[b] My addition.

[c] Presumably classes and hierarchies of previous eras are here replaced by inter-penetrating, functional, societal networks.

To be able to contend with and comprehend what is happening seems to demand a much more flexible, combinatorial, and cautiously eclectic specialization of thought and action, theory and practice, than the old modernist intellectual division of labor, with its distinctive positioning of history and geography, would allow. Flexible specialization, resistance to paradigmatic closure and to rigidly categorical thinking, the capacity to combine effectively what in the past was considered antithetical/uncombinable, the selective rejection of all embracing 'deep logics', the search for new ways to interpret the empirical world have become hallmarks of an emerging postmodern perspective.

He also sees post-modern human geography as a perspective on geographically uneven development in which the differentiation of landscapes is 'not only a social product but also rebounds back to shape social relations, social practice, social life itself. . . . "socio-spatial dialectic" ' (ibid. 291; see also Soja, 1980).

Recently, the 'group planners' of a large multinational company sought out the values of the present front runners of socio-cultural change in western Europe, 'i.e. the people who are ahead in the trends that are gaining significance in society at large'. Here is the answer:

The frontrunners are not either rational or emotional, but combine rational analysis with an intuitive approach. More importantly, they cope easily with complexity and ambiguity. They tend to like complexity, and dislike situations where everything is spelled out for them. As a result, they could be called 'strategic opportunists': they work hard in a job or for a project but, when no further progress seems to be made [based on some external pragmatic test?], they change course and do something else with the same degree of enthusiasm. They can change . . . motivations, . . . see no reason to be 'consistent' . . . play different roles depending on circumstances, and the overall pattern is likely to be different in the various stages of their lives. (de Vries, 1988, 3–4)

Here we may have the description of the current, trend-setting generation of geographers and a hint of the kind of geography they will produce for the nineties.

Turning back to the present, I see more intellectual room now in the subject of geography than when I began my studies. That is what I call progress to something better, not because we are now nearer any ultimate truths, but because there are now more opportunities for those 'stimulating comparisons' between the ideas that we hold, always on probation. A vigorous discipline like modern geography is a rough sea where judgement based on experience is needed for navigation. The reward for voyaging over the open ocean is that there are always new horizons to traverse. 'We can break out of our framework at any time.' Yes, another quotation from Popper (1970, 56), but I should like to give the last word to one whom Popper himself has called his 'eminent predecessor' who 'anticipated several of my ideas—ideas I had independently' (Popper, 1974, 1119). This was Charles Sanders Peirce, who managed to conceive of a libertarian, open-ended view of science, even though he himself was embedded in the time structure of the nineteenth century, and first published the following seven years before Einstein began to topple Newton-inspired certainties:

Nothing is *vital* for science; nothing can be. Its accepted propositions, therefore, are but opinions at most; and the whole list is provisional. The scientific man is not in the least wedded to his conclusions. He risks nothing upon them. He stands ready to abandon one or all as soon as experience opposes them. (Peirce, 1931–5, i. 347, originally 1898)

Appendix I

Relevant active debates in philosophy

I: Dominant bases of philosophy and their relevance to geography

A recent symposium focuses on philosophy's 'interpretive turn', succeeding phases when that discipline was dominated by epistemological and linguistic preoccupations (see Table 40). Many geographers still rely on epistemological–ontological foundations, unaffected by philosophy's later 'turns'. It is certainly difficult to see much direct influence of philosophy's language debates upon geographical theorizing, but the current 'interpretive turn' is of prime interest. This is because the focus of the debate is upon the methodologies of the natural and human sciences and whether or not there is fundamental difference between them (Table 35 and pp. 231–5).

There are philosophers (notably Rorty, see this Appendix, Part II) who argue that a belief in a basic methodological difference between the natural and human sciences has outlived its usefulness. Others who believe in such a distinction disagree on where to draw the line. The 'Rorty school' asserts that we are all interpreters now; the other holds that there is some objective reality (see again this Appendix, Part II) for the natural sciences to investigate whereas the human sciences operate under a 'double hermeneutic' (see pp. 80–1 and Table 40).

Obviously, I cannot go on here to discuss this philosophical debate in all its complexity, but observation of its future course will throw interesting light on the fundamental relationship between physical and human geography within the subject as a whole.

II: The 'real world' or can I get away from and beyond 'That's what I think'?

A review (Mason, 1991, 89) of the first edition complains that there is no mention of Rorty (1979, his major work), and a recent

TABLE 40.   Dominant bases of philosophy described as 'turns'

Turn	Dates	Major characteristics	Associated features
Epistemological	1700 to early in this century	Philosophy provides an epistemological–ontological basis for enquiry	Distinction between explanation and interpretation; and between natural and human sciences
Linguistic	Most of this century	Preoccupations with the structure of language, word–world relationships, and the analysis of meaning	Distinction between conceptual and empirical; and between scientific enterprise and interpretive disciplines
Interpretive	Recent to date	Importance of interpretation in *all* enquiry	Debate: are methods in the natural and human sciences the same? Or, are they different? (because the human sciences are 'doubly hermeneutic' — 'They do not just give interpretations, they are interpretations of interpretations.' (p. 5.)

Source: Hiley, Bohman, and Shusterman (1991, 1–14).

compendium of 'critical responses' to Rorty's arguments (Malachowski, 1990) makes the complaint appear justified. For we can now see that the question of the claims about the existence or otherwise of the real world is still the subject of active debate in philosophy. I did mention this debate (p. 196 above) and quoted arguments for the existence of the real world by Popper (Popper and Eccles 1977, 10) and James (1920, 67; originally 1878). Malachowski (1990, 139) has now provided three comments on this question by Kant, Russell, and Heidegger:

it still remains a scandal to philosophy and to human reason in general that the existence of things outside us . . . must be accepted on *faith*, and that if anyone thinks good to doubt their existence, we are unable to counter his doubts by a satisfactory proof. (Kant, 1781, in 1978 trans., preface to 2nd edn., 34)

If you are willing to believe that nothing exists except what you directly experience, no other person can prove you wrong, and probably no valid arguments against your view exist. (Russell, 1923)

Heidegger takes a contrary view:

The 'scandal of philosophy' is not that this proof has yet to be accepted but that *such proofs are expected and attempted again and again*. (1927, in trans. 1962, 249)

Rorty has reactivated this debate, coming down on Heidegger's 'side', and we can use the following summary of his basic position offered by Sorell (1990, 11):

Rorty not only denies that the scientist puts us in touch with a transcendental intrinsic nature of the world; he doubts whether there is an intrinsic nature of the world waiting to be discovered by science. He thinks that the idea of the world's intrinsic nature takes for granted an outmoded conception of the world as God's artefact, as 'the work of someone who has something in mind' . . . [Rorty 1986, 6]

Two of the 'critical responses' to Rorty in Malachowski's edited volume of 1990 are by Williams (1990) and Taylor (1990). Part of their basic argument is neatly summarized in a footnote by Taylor (1990, 274 n. 14):

I think that Bernard Williams has made a point very similar to mine very tellingly, when he says . . . (. . . 1985 [, 137–8]) . . . that Rorty's account 'is self-defeating. If the story he tells were true, there would be no perspective from which he could express it'. In saying that the best scientific language is best not in virtue of describing a world already there, but in virtue of its being convenient or advantageous for us to adopt, 'he is trying to reoccupy the transcendental standpoint outside human speech and activity, which is precisely what he wants us to renounce'.

My own view is that there does exist a real world out there (Popper's World 1; this has to be considered alongside his World 2, that which is inside our individual heads; and World 3, objective knowledge; see further in pp. 199–204, above). This of course leaves me open to the charge that my 'real world out there' is, on my own confession, part of my 'own view', or perhaps more bluntly: 'Huh! that's what you think!'—a criticism which is somewhat of a mirror image to that used by Taylor and Williams against Rorty. In this paragraph we have the essence of the debate in philosophy about the reality of the world which has extended from Descartes to Rorty, and now beyond.

But to return to 'My own view': it is not only based on just what I think, on an unsupported epistemology, but on the evidence of what is found in practice, in a mutually supporting methodology-epistemology. And the type of evidence on which I rely can be found in the arguments already cited, on p. 196 above, by Popper (Popper and Eccles 1977, 10) and James (1920, 67; originally 1878).

Appendix II

Points of view; and the great 'As if'

There have been several recent general surveys of wide subject areas within geography, and each author, including myself, has encountered the same problem: the more coherent the point of view from which the survey is made, the more criticism may be attracted from those with other points of view. The great nine-teenth-century novelists often adopted a neutral, sometimes faintly ironic, perspective as they described scenes and narrated events. They had created their entire world and told us as much about it as they thought artistically fit. When we come to surveys of academic disciplines which contain widely contrasting points of view, we can of course either adopt one of the existing perspec-tives or offer something novel. We cannot be neutral or outside the field of action in which we are partisan players by virtue of our very profession. Where does our point of view come from? I have tried to show that it cannot be based on the a priori foundation of an epistemology because of the diallelus problem (pp. 225–6). If the point of view is based on some conviction, the question imme-diately arises: 'What is the evidential basis for the conviction?' We here encounter a fundamental opposition between dogmatic (unchanging) and critical (mutating) thinking, which occurs in Popper's *Objective Knowledge* (1972, 347), but here I should like to quote him from a little-known paper (1976*b*, 1): 'From a biological point of view, the theories or hypotheses of the empirical sciences can be regarded as attempted biological adaptations: they are analogous to mutations, or to tentative anticipations, or to expec-tations. . . . The critical scientist actively tries to eliminate his false theory. He tries to let his false theory die in his stead. The dogmatic believer—whether animal or man—often perishes with his false beliefs.' So our point of view must have an inbuilt capacity to mutate.

This was one of the reasons that led me to suggest a pragmatic analytical methodology–epistemology (PAME) which is my point of view in Step 3 (below) of a simple exposition of how to settle on a perspective:

1. Survey the problem field to get a feel of its important headings. A 'problem' here is the inevitability of a point of view.
2. Have a hunch as to what might be a successful point of view. One attribute of this must be the perspective's 'inbuilt capacity to mutate'.
3. Test the putative perspective for effectiveness in the relevant field, and refine it as far as possible.
4. Let posterity judge whether or not the suggestion in Step 3 achieves pragmatic success (usefulness) or otherwise.

I cannot see that adherence to a particular point of view (on probation) should completely preclude contrasting snapshots or vignettes from other points of view, perhaps as quotations, to give illumination by highlights of contrast; and I have cited the 'humanistic' verbatim quotations from interviewees alongside a hypothetico-deductive analysis of questionnaire returns (p. 239).

Most general accounts of modern human geography rail against positivism (often erroneously held to be synonymous with scientific method), citing as one of its faults the claim to be objectively outside the subject studied, and a survey by Cloke, Philo, and Sadler (1991, 14) is no exception: 'In the first instance it [positivism] arguably creates a false sense of objectivity by artificially separating the observer from the observed . . .'. Yet in the recent past general commentators on geography have variously claimed that they are neutral observers, or believers in seeing geography (from 'outside' or 'above') as pluralistic. Cloke, Philo, and Sadler (1991, 205) argue for 'a kind of relativism that accepts the incompatibility of different approaches in human geography . . . and that still strives to allow incompatibles to enter into constructive (rather than destructive) dialogue'. But we are not just auditors of dialogue. We take part. Even as we speak and write we have some frame of reference and must acknowledge what this is, so that our hearers and readers can be on their guard against what *their* perspective tells them are our biases.

You may complain of a discrepancy between this stated impossibility of standing outside debates and the attempt to find evidence for the real world out there (World 1; see Appendix I).

Fine (1984; see also Boyd, Gasper, and Trout, 1991) has studied this question in some depth, and a little selective quotation from him will illustrate my argument:

we are in the world, both physically and conceptually. That is, *we* are among the objects of science, and the concepts and procedures that we use to make judgments of subject matter and correct application are themselves part of that same scientific world. (p. 272)

If you believe that guessing based on some truths [e.g. the results we get are those we would expect if the real world exists] is more likely to succeed than guessing pure and simple, then if our earlier theories were in large part true and if our refinements of them conserve the true parts, then guessing on this basis has some relative likelihood of success. (p. 273)

This argument for 'scientific postrealism'* is modestly put, but I like the reference to a pragmatic test in the last three words. Now if your complaint is that my position involves having my real cake 'out there' and of also mentally digesting it, then I can only say that you have understood an essential part of PAME.

Finally, we arrive at the question of the random element in the universe, both in physical and human geography, where 'the best laid plans . . .'. Some problems may not be solvable by current rational processes at the time decision is required; political compromise agreed by a majority is necessary. Other problems may change as we attempt to solve them, or even *because* we attempt to solve them. Randomness may be a critical element in the problem data. Do we therefore throw up our hands as some extreme postmodernists may do in their enthusiasm to reject totalizing, or even generalizing discourses? We can all cite the experience of many individuals whose lives have been vitally affected by blind chance. But this does not happen to all of us for all of the time. Meanwhile we follow the belief in the great 'As if': we act as if the universe

* In this context 'scientific realism' is rejected, having been defined as: 'The view that the subject matter of scientific research and scientific theories exists independently of our knowledge of it, and that the goal of science is the description and explanation of both the observable and unobservable aspects of an independently existing world' (Boyd, Gasper, and Trout, 1991, 760). Relativism is also rejected when defined as follows: 'In epistemology, the view that the acceptability or unacceptability of knowledge claims is relative to a particular group or community, and that there are no objective epistemological standards' (ibid.).

were rational, and this pays off for a livable percentage of the time. Just think of those few magic moments when a great sporting occasion or theatrical performance is about to begin; then think of the thousands of individual decisions and plans that have been successfully carried out so that everyone, participants and audience, is ready at the same time. Our belief in the ability to solve problems by planning ahead is only temporarily shaken by an unlucky random accident.

An illustration of the contrast between attitudes to order and randomness is provided by a strange episode in Hammett (1974) *The Maltese falcon* and may be briefly summarized as follows. A married man named Flitcraft who has an orderly existence just avoids being hit by a falling beam from a building site.

The life he knew was a clean orderly sane responsible affair. Now a falling beam had shown him that life was fundamentally none of these things. . . . What disturbed him was the discovery that in sensibly ordering his affairs he had got out of step, and not into step, with life. [So he decides to] change his life at random by simply going away. (Hammett, 1974, 56)

He wanders aimlessly for a couple of years and then settles down again, under a new name, marries another woman, and moves into another ordered existence.

Marcus (1974, pp. xviii–xix) has commented on this episode to reveal a fundamental paradox.

For Flitcraft the falling beam 'had taken the lid off life and let him look at the works'. [Hammett, 1974, 56] The works are that life is inscrutable, opaque, irresponsible, and arbitrary—that human existence does not correspond in its actuality to the way we live it. For most of us live as if existence itself were ordered, ethical, and rational. As a direct result of his realization in experience that it is not, Flitcraft leaves his wife and children and goes off. He acts irrationally and at random, in accordance with the nature of existence. When after a couple of years of wandering aimlessly about he decides to establish a new life, he simply reproduces the old one he had supposedly repudiated and abandoned; that is to say, he behaves again as if life were orderly, meaningful, and rational, and 'adjusts' to it. . . . here we come upon the unfathomable and most mysteriously irrational part of it all—how despite everything we have learned and everything we know, men will persist in behaving and trying to behave sanely, rationally, sensibly, and responsibly. And we will continue to persist even when we know that there is no logical or

metaphysical, no discoverable or demonstrable reason for doing so.*
The contradiction . . . is not merely sustained; it is sustained with pleasure. For Hammett and Spade [Sam Spade, Hammett's detective hero]
. . . the sustainment in consciousness of such contradictions is an indispensable part of their existence and of their pleasure in that existence.

Let me repeat a sentence from the above quotation, for it
describes a powerful factor that we must reckon with: 'For most of
us live as if existence itself were ordered, ethical and rational.'
This 'As if' is a fundamental paradox of the way we live and also
lies within the critical rationalist hypothetico-deductive version of
the scientific method where we study the world *as if* it were
ordered and rational, writing about it, we hope, in ordered and
rational fashion. But in doing so we always find that there are 'bits
that don't fit'. If the bits add up to a sharp reef of antagonism confronting our method of navigating, we must try another tack until
the residual rocks are reduced to proportions that fail to sink our
thesis. These residuals may be entirely resistant to any version of
the scientific method, which is where humanistic methods come
into their own. But these 'odd bits' of today, far from being forever
unimportant, may contain the seeds of the important problems
that challenge us tomorrow.

* It can hardly be an accident that the new name that Hammett gives to
Flitcraft [Charles Pierce] is that of an American philosopher—with two vowels
reversed—who was deeply involved with such speculations.

REFERENCES

Adams, R.M. 1960*a*: The origin of cities, *Scientific American*, Sept. Freeman: Scientific American Resource Library; offprint, No. 606.

—— 1960*b*: Factors influencing the rise of civilization in the alluvium: illustrated by Mesopotamia. In C.H. Kraeling and R.M. Adams (eds.), *City invincible*. Chicago: University of Chicago Press, 24–34.

—— 1966: *The evolution of urban society: early Mesopotamia and prehispanic Mexico*. Chicago: Aldine.

Agassi, J. 1966: The confusion between science and technology in the standard philosophies of science, *Technology and Culture*, 7, 348–66.

Agate, J. 1928: What is a good play? *Radio Times*, 7 Dec.

Agnew, C.T. 1984: Checkland's soft systems approach: a methodology for geographers, *Area*, 16, 167–74.

Agnew, J.A. and Duncan, J.S. 1981: The transfer of ideas into Anglo-American human geography, *Progress in Human Geography*, 5, 42–57.

Alchian, A. 1950: Uncertainty, evolution, and economic theory, *Journal of Political Economy*, 58, 211–21.

Allee, W.C. 1934: Concerning the organization of marine coastal communities, *Ecological Monographs*, 4, 541–54.

Allen, J. 1983: Property relations and landlordism: a realist approach, *Environment and Planning D: Society and Space*, 1, 191–203.

—— 1987: Realism as method, *Antipode*, 19, 231–9.

—— 1988: Realism in geography: use and limitations. Paper delivered to the Annual Conference, Institute of British Geographers, Loughborough, 7 Jan.

Althusser, L. and Balibar, E. 1970: *Reading capital*. London: New Left Books.

Ambrose, J. and Williams, C.H., 1981: *Scale as an influence on the geolinguistic analysis of a minority language*, Discussion Papers in Geolinguistics, 4. Stafford: North Staffordshire Polytechnic, Department of Geography and Sociology.

Andrews, J.T. 1972: Glacier power, mass balances, velocities, and erosion potential. In A. Cailleux (ed.), *Glacial and periglacial morphology*. Zeitschrift für Geomorphologie Supplementband 13. Berlin and Stuttgart: Borntraeger, 1–17.

Anuchin, K.A. 1977: *Theoretical problems of geography*, ed. R. Fuchs and G. Demko. Trans; first pub. 1960; Columbus: Ohio University Press.

Appleton, J. 1975: *The experience of landscape*. London: Wiley.

Archer, K. 1987: Mythology and the problem of reading in urban and regional research, *Environment and Planning D: Society and Space*, 5, 384–93.

Ardrey, R. 1967: *The territorial imperative: a personal inquiry into the animal origins of property and nations*. London: Collins.

Arieti, S. 1976: *Creativity: the magic synthesis*. New York: Basic Books.

Ashby, W.R. 1956: *Introduction to cybernetics*. London: Wiley.

Bain, J.S. 1954: Economies of scale, concentration and the conditions of entry in twenty manufacturing industries, *American Economic Review*, 44, 15–39.

Baker, A.R.H. 1981: On ideology and historical geography. In A.R.H. Baker and M. Billinge (eds.), *Period and place: research methods in historical geography*. Cambridge: Cambridge University Press.

Bale, J. and McPartland, M. 1986: Johnstonian anarchy, inspectorial interest and undergraduate education of PGCE students, *Journal of Geography in Higher Education*, 10, 61–70.

Ball, M. 1987: Harvey's Marxism, *Environment and Planning D: Society and Space*, 5, 393–4.

Barbour, I.G. 1974: *Myths, models and paradigms: the nature of scientific and religious language*. London: SCM Press.

Barnes, T. and Curry, M. 1988: Metaphor, postmodernism, and economic geography, *Association of American Geographers Annual Meeting, Program and Abstracts*, 6–10. Apr. Phoenix, 9.

Barthes, R. 1968: *Elements of semiology*: Trans. A. Lavers and C. Smith; London: Cape.

Bartlett, F.C. 1928: Types of imagination, *Journal of Philosophical Studies*, 3, 78–85.

—— 1932: *Remembering: a study in experimental and social psychology*. Cambridge: Cambridge University Press.

—— 1958: *Thinking: an experimental and social study*. London: Allen and Unwin.

Barton, R. 1980, 1983: *Radical geography: a research bibliography*. Working Papers, Nos. 291 and 357. Leeds: University of Leeds, Department of Geography.

Barzun, J. 1974: *Clio and the doctors: psycho-history, quanto-history and history*. Chicago: University of Chicago Press.

Baynes, K., Bohman, J., and McCarthy, T. (eds.) 1987: Introduction to J.F. Lyotard, The postmodern condition, *After philosophy, end or transformation?* Cambridge, Mass.: MIT Press, 67–72.

Beer, S. 1977: Cybernetics. In A. Bullock and O. Stallybrass (eds.), *The Fontana dictionary of modern thought*. London: Fontana, 151.

Bennett, R.J. and Chorley, R.J. 1978: *Environmental systems: philosophy, analysis and control*. London: Methuen.

Berdoulay, V. 1978: The Vidal–Durkheim debate. In D. Ley and M. Samuels (eds.), *Humanistic geography: prospects and problems*. London: Croom Helm, 77–90.

Berger, P. and Pullberg, S. 1965: Reification and the sociological critique of consciousness, *History and Theory*, 4, 196–211.

—— and Luckmann, T. 1966: *The social construction of reality*. Harmondsworth: Penguin.

Bergin, T.G. and Fisch, M.H. 1961: *The new science of Giambattista Vico*. First pub. 1744; New York: Anchor.

Bergmann, G. 1957: *Philosophy of science*. Madison: University of Wisconsin Press.

Berlin, I. 1973: The counter-enlightenment. In P.P. Wiener (ed.), *Dictionary of the history of ideas*, New York: Scribner, ii. 100–12.

Bernstein, R.J., 1966: *John Dewey*. New York: Washington Square.

Berry, B. J. L. 1959: Ribbon developments in the urban business pattern, *Annals of the Association of American Geographers*, 49, 149–55.

—— 1963: *Commercial structure and commercial blight*, Department of Geography Research Paper, No. 85. Chicago: University of Chicago.

—— 1964: Approaches to regional synthesis, *Annals of the Association of American Geographers*, 54, 2–11.

—— and Horton, F. E. 1970: *Geographic perspectives on urban systems.* Englewood Cliffs, NJ: Prentice-Hall.

Bhaskar, R. 1975: *A realist theory of science.* York: Alma.

Bierce, A. L. 1963: A lacking factor. In G. Barkin (ed.), *The sardonic humor of Ambrose Bierce.* Selected from the collected works of Ambrose Bierce pub. 1909–1912; New York: Dover.

Billinge, M. 1977: In search of negativism: phenomenology in historical geography, *Journal of Historical Geography*, 3, 55–67.

Birch, B. P. 1981: An English approach to the American frontier, *Journal of Historical Geography*, 7, 397–406.

Bird, J. 1956: Scale in regional study illustrated by brief comparisons between the western peninsulas of England and France, *Geography*, 41, 25–38.

—— 1957: *The geography of the port of London.* London: Hutchinson.

—— 1963: The noosphere: a concept possibly useful to geographers, *Scottish Geographical Magazine*, 79, 54–6.

—— 1967: Seaports and the European Economic Community, *Geographical Journal*, 133, 302–27.

—— 1971: *Seaports and seaport terminals*, London: Hutchinson.

—— 1973: Desiderata for a definition: or is geography what geographers do? *Area*, 6, 201–3.

—— 1975: Methodological implications for geography from the philosophy of K. R. Popper, *Scottish Geographical Magazine*, 91, 153–63.

—— 1977*a*: *Centrality and cities.* London: Routledge and Kegan Paul.

—— 1977*b*: Methodology and philosophy, *Progress in Human Geography*, 1, 104–10.

—— 1978: Methodology and philosophy, *Progress in Human Geography*, 2, 133–40.

—— 1979: Methodology and philosophy, *Progress in Human Geography*, 3, 117–25.

—— 1980: Seaports as a subset of gateways for regions: a research survey, *Progress in Human Geography*, 3, 360–70.

—— 1981: The target of space and the arrow of time, *Transactions of the Institute of British Geographers*, 6, 129–51.

—— 1982: Transport decision-makers speak: the Seaport Development in the European Communities Research Project, *Maritime Policy and Management*, 9, pt. I: 1–22, pt. II: 83–102.

—— 1983*a*: Transactions of ideas: a subjective survey of the Transactions during the first fifty years of the Institute, *Transactions of the Institute of British Geographers*, 8, 55–69.

—— 1983*b*: Gateways: slow recognition but irresistible rise, *Tijdschrift voor Economische en Sociale Geografie*, 74, 196–201.

—— 1985: Geography in three worlds: how Popper's system can help elucidate dichotomies and changes in the discipline, *The Professional Geographer*, 37, 403–9.

— 1986: Gateways: examples from Australia, with special reference to Canberra, *Geographical Journal*, 152, 56–64.

— 1988: Freight forwarders speak: the Perception of Route Competition via Seaports in the European Communities Research Project, pt. II, *Maritime Policy and Management*, 15, 107–25.

— and Bland, G. 1988: Freight forwarders speak: the Perception of Route Competition via Seaports in the European Communities Research Project, pt. I, *Maritime Policy and Management*, 15, 35–55.

— and Pollock, E. E. 1978: The future of seaports in the European Communities, *Geographical Journal*, 144, 23–48.

— and Witherick, M. E. 1986: Marks and Spencer: the geography of an image, *Geography*, 71, 305–19.

— Lochhead, E. N. and Willingale, M. C. 1983: Methods of investigating decisions involving spatial effects including content analysis of interviews, *Transactions of the Institute of British Geographers*, 8, 143–57.

Bishop, A. 1978: Human geography: a nomothetic science? *Area*, 10, 149–51.

Blizzard, S. W. and Anderson, W. F. 1952: *Problems in rural–urban fringe research: conceptualisation and delineation*. Progress Report, No. 89. University Park, Pa: Pennsylvania Agricultural Experiment Station State College.

Block, R. H. 1980: Frederick Jackson Turner and American geography, *Annals of the Association of American Geographers*, 70, 31–42.

Bonjour, L. 1979: Rescher's epistemological system. In E. Sosa (ed.), *The philosophy of Nicholas Rescher: discussion and replies*. Dordrecht: Reidel, 157–72.

Boulding, K. 1956: General system theory: the skeleton of science, *Yearbook of the Society of Advanced General Systems Theory*, 1, 1–17.

Bourne, L. E., Dominowski, R. L., and Loftus, E. F. 1979: *Cognitive processes*. Englewood Cliffs, NJ: Prentice-Hall.

Bourne, L. S. 1974: Urban systems in Australia and Canada: comparative notes and research questions, *Australian Geographical Studies*, 12, 152–72.

Boyd, R., Gasper, P., and Trout, J. D. (eds.) 1991: *The philosophy of science*. Cambridge, Mass.: MIT.

Bragaw, L. K., Marcus, H. S., Raffaele, G. C., and Townley, J. R. 1975: *The challenge of deepwater terminals*. Lexington, Mass.: Lexington.

Braudel, F. 1972: *The Mediterranean and the Mediterranean world in the age of Philip II*, 2 vols. London: Collins.

Brown, L. A. 1981: *Innovation diffusion: a new perspective*. London: Methuen.

Bruton, H. J. 1960: Contemporary theorizing on economic growth. In B. F. Hoselitz, J. J. Spengler, J. M. Letiche, E. McKinley, J. Buttrick and H. J. Bruton (eds.), *Theories of economic growth*. Glencoe, Ill.: Free Press, 239–98.

Büdel, J. 1982: *Climatic geomorphology*. Princeton, NJ: Princeton University Press.

Bull, W. B. 1975: Allometric change of landforms, *Geological Society of America Bulletin*, 86, 1489–98.

Bunge, W. 1958: *The location of population demand, purchasing power, and services and highways.* Seattle: University of Washington, Department of Geography.

—— 1966a: *Theoretical geography.* 2nd edn., Lund: Gleerup.

—— 1966b: Annals commentary: locations are not unique, *Annals of the Association of American Geographers,* 56, 375–6.

—— 1974: Regions are sort of unique, *Area,* 6, 92–9.

Bunting, T. E. and Guelke, L. 1979: Behavioral and perception geography: a critical appraisal, *Annals of the Association of American Geographers,* 69, 448–62 (with ensuing discussion, 463–74).

Burgess, R. 1976: *Marxism and geography,* Occasional Paper, No. 30. London: University College London, Department of Geography.

Burrill, M. F. 1968: The language of geography, *Annals of the Association of American Geographers,* 58, 1–11.

Burton, I. 1963: The quantitative revolution and theoretical geography, *The Canadian Geographer,* 7, 151–62.

Butlin, R. 1987: Theory and methodology in historical geography, In M. Pacione (ed.), *Historical geography: progress and prospect.* London: Croom Helm, 16–45.

Buttimer, A. 1976: Grasping the dynamism of lifeworld, *Annals of the Association of American Geographers,* 66, 277–92.

—— 1978: Charisma and context: the challenge of La géographie humaine. In D. Ley and M. Samuels (eds.), *Human geography: prospects and problems.* London: Croom Helm, 58–76.

—— 1979: Reason, rationality, and human creativity, *Geografiska Annaler,* 61B, 43–9.

—— 1983: *The practice of geography.* London: Longman.

—— and Seamon, D. (eds.) 1980: *The human experience of space and time.* London: Croom Helm.

Catanese, A. J. 1972: *Scientific methods of urban analyses.* Urbana, Ill.: University of Illinois Press.

Cesario, F. J. 1975: A primer on entropy modelling, *Journal of the American Institute of Planners,* 41, 40–8.

Chalmers, A. F. 1978: *What is this thing called science?* Milton Keynes: Open University.

Champion, A. G. and Green, A. 1987: The booming towns of Britain: the geography of economic performance in the 1980s, *Geography,* 72, 97–108.

Chapman, G. P. 1977: *Human and environmental systems: a geographer's appraisal.* London: Academic Press.

Chatterjee, L., Harvey, D., and Klugman, L. 1974: *FHA policies and the Baltimore City Housing Market.* Baltimore: Baltimore Urban Observatory.

Checkland, P. B. 1972: Towards a systems-based methodology for real-world problem solving, *Journal of Systems Engineering,* 3, 87–117.

—— 1981: *Systems thinking, systems practice.* Chichester: Wiley.

Chesterton, G. K. 1904: *The Napoleon of Notting Hill.* London: Bodley Head.

Chisholm, M. 1960: The geography of commuting, *Annals of the Association of American Geographers,* 50, 187–8 and 491–2.

—— 1967: General systems theory and geography, *Transactions of the Institute of British Geographers,* 42, 45–52.

—— 1979: Von Thünen anticipated, *Area*, 11, 37–9.

Chorley, R.J. 1962: Geomorphology and general systems theory, *US Geological Survey Professional Paper*, 500-B, 1–10.

—— 1973: Geography as human ecology, In R.J. Chorley (ed.), *Directions in geography*. London: Methuen, 155–69.

—— 1976: Some thoughts on the development of geography from 1950 to 1975. In D. Pepper and A. Jenkins (eds.), *Proceedings of the 1975 National Conference on Geography in Higher Education*. Oxford Polytechnic Discussion Paper in Geography, No. 3. Oxford: Oxford Polytechnic Press.

—— 1987: Perspectives on the hydrosphere. In M.J. Clark, K.J. Gregory, and A.M. Gurnell (eds.), *Horizons in physical geography*. London: Macmillan, 378–81.

—— and Haggett, P. (eds.) 1967: *Models in geography*. London: Methuen.

—— and Kennedy, B. 1971: *Physical geography: a systems approach*. London: Prentice-Hall.

Chouinard, V., Fincher, R., and Webber, M. 1984: Empirical research in scientific human geography, *Progress in Human Geography*, 8, 347–80.

Church, M. and Mark, D.M. 1980: On size and scale in geomorphology, *Progress in Physical Geography*, 4, 342–90.

Clark, D., Davies, W.K.D., and Johnston, R.J. 1974: The application of factor analysis in human geography, *The Statistician*, 23, 259–81.

Clark, G.L. and Dear, M. 1978: The future of radical geography, *The Professional Geographer*, 30, 356–9.

Clark, K.G.T. 1950: Certain underpinnings of our arguments in human geography, *Transactions of the Institute of British Geographers*, 16, 13–22.

Clark, M.J. (ed.) 1988: *Advances in periglacial geomorphology*. Chichester: Wiley.

—— Gregory, K.J., and Gurnell, A.M. 1987: Physical geography: diversity and unity: a concluding perspective. In M.J. Clark, K.J. Gregory, and A.M. Gurnell (eds.), *Horizons in physical geography*. London: Macmillan, 382–6.

Clark, R. 1974: Defining and doing, or doing and knowing, *Area*, 6, 154–7.

Clarkson, J.D. 1970: Ecology and spatial analysis, *Annals of the Association of American Geographers*, 60, 700–16.

Claval, P. 1967: Qu'est-ce que la géographie? *Geographical Journal*, 133, 33–9.

—— 1984: Conclusion. In R.J. Johnston and P. Claval (eds.), *Geography since the Second World War: an international survey*. London: Croom Helm, 282–9.

Cloke, P., Philo, C., and Sadler, D. 1991: *Approaching human geography*. London: Chapman.

Clout, H., Blacksell, M., King, R., and Pinder, D. 1985: *Western Europe: geographical perspectives*. London: Longman.

Cochrane, A. 1987: What a difference the place makes: the new structuralism of locality, *Antipode*, 19, 354–63.

Coffey, W.J. 1981: *Geography: towards a general spatial systems approach*. London: Methuen.

Colby, C.C. 1933: Centrifugal and centripetal forces in urban geography, *Annals of the Association of American Geographers*, 23, 1–20.

Cole, J.P. 1986: *The poverty of Marxism in contemporary geographical applications and research: the search for the Emperor's new clothes*, an extended discussion paper. Nottingham: University of Nottingham, Department of Geography.

— and King, C. A. M. 1968: *Quantitative geography*. London: Wiley.

Coleman, A. 1985: *Utopia on trial*. London: Shipman.

— 1987: Utopia on trial: a comment on P. Dickens's review, *International Journal of Urban and Regional Research*, 11, 115–17.

Collingwood, R. G. 1946: *The idea of history*. Oxford: Oxford University Press.

Conway, D. 1987: *A farewell to Marx: an outline and appraisal of his theories*. Harmondsworth: Penguin.

Cooke, P. 1987: Clinical inference and geographic theory, *Antipode*, 19, 69–78.

Corbridge, S. 1986: *Capitalist world development: a critique of radical development geography*. London: Macmillan.

Cosgrove, D. 1983: Review of L. Guelke, *Historical understanding in geography: an idealist approach*, *Landscape History*, 5, 100–1.

— and Thornes, J. E. 1981: Of truth of clouds: John Ruskin and the moral order in landscape. In D. C. D. Pocock (ed.), *Humanistic geography and literature: essays on the experience of place*. London: Croom Helm, 20–46.

Costall, A. 1981: On how so much information controls so much behaviour: James Gibson's theory of direct perception. In G. Butterworth (ed.), *Infancy and epistemology*. Brighton: Harvester, 30–51.

— 1986: The 'psychologist's fallacy' in ecological realism, *Teorie & Modelli*, 3, 37–46.

Couclelis, H. 1986: A theoretical framework for alternative models of spatial decision and behavior, *Annals of the Association of American Geographers*, 76, 95–113.

— and Golledge, R. G. 1983: Analytic research, positivism, and behavioral geography, *Annals of the Association of American Geographers*, 73, 331–9.

Cox, K. R. 1981: Bourgeois thought and the behavioral geography debate. In K. R. Cox and R. G. Golledge (eds.), *Behavioral problems in geography revisited*. London: Methuen, 256–79.

Crist, E. P. and Kauth, R. J. 1986: The tasseled cap de-mystified, *Photogrammetric Engineering and Remote Sensing*, 52, 81–6.

Curry, M. 1986: On possible worlds from geographies of the future to future geographies. In L. Guelke (ed.), *Geography and humanistic knowledge*. Waterloo Lectures in Geography, ii. Publication Series No. 25; Waterloo: University of Waterloo, Department of Geography, 87–99.

Daiches, D. and Flower, J. 1979: *Literary landscapes of the British Isles: a narrative atlas*. London: Paddington Press.

Daniels, S. 1984: Review of L. Guelke, *Historical understanding in geography: an idealist approach*, *Journal of Historical Geography*, 10, 225–6.

Darby, H. C. 1962a: The problem of geographical description, *Transactions of the Institute of British Geographers*, 30, 1–14.

— 1962b: Historical geography. In H. P. E. Finberg (ed.), *Approaches to history: a symposium*. London: Routledge and Kegan Paul, 127–56.

Darling, F. F. 1952: Social behaviour and survival, *Auk*, 69, 183–91.

Davies, W. K. D. 1972: Geography and the methods of modern science. In W. K. D. Davies (ed.), *The conceptual revolution in geography*. London: University of London Press, 131–9.

Davis, W. M. 1899: The geographical cycle, *Geographical Journal*, 14, 481–504.

Day, M. D. 1980: Dialectical materialism and geography, *Area*, 12, 142–6.

Dear, M. J. 1986: Postmodernism and planning, *Environment and Planning D: Society and Space*, 4, 367–84.

—— 1987: Editorial: Society, politics, and social theory, *Environment and Planning D: Society and Space*, 5, 363–6.

—— 1988: The postmodern challenge: reconstructing human geography, *Transactions of the Institute of British Geographers*, 13, 262–74.

—— Jackson, P., Thrift, N. J., and Williams, P. R. 1987: Cities, consumption, culture, and postmodernism: books in 1986, *Environment and Planning D: Society and Space*, 5, 475–84.

Desbarats, J. 1983: Spatial choice and constraints on behavior, *Annals of the Association of American Geographers*, 73, 340–57.

De Vries, J. 1988: *Images of the nineties*. London: Shell International Petroleum Company, Group Public Affairs.

Dewey, J. 1929: *The quest for certainty*. New York: Minton, Balch.

—— 1938: *Logic: the theory of inquiry*. New York: Holt.

—— 1960: Qualitative thought. In R. J. Bernstein (ed.), *On experience, nature and freedom: representative selections: John Dewey*. First pub. 1930; New York: Liberal Arts, 176–98.

—— 1976: *Lectures on psychological and political ethics*, ed. D. F. Koch. First pub. 1898; New York: Hafner.

Dickens, P. 1986: Review of A. Coleman, *Utopia on trial*, and E. Krupat, *People in cities: the urban environment and its effects*, *International Journal of Urban and Regional Research*, 10, 297–300.

—— 1987: Utopia on trial: a response to Alice Coleman's comment, *International Journal of Urban and Regional Research*, 11, 118–20.

Dilthey, W. 1957: *Gesammelte Schriften*, v. 2nd edn., ed. G. Mirsch. First pub. 1924; Stuttgart: Teubner.

Douglas, I. 1986: The unity of geography is obvious . . .,*Transactions of the Institute of British Geographers*, 11, 459–63.

Downs, R. M. 1970: Geographic space perception: past approaches and future prospects, *Progress in Geography*, 2, 65–108.

—— and Stea, D. 1977: *Maps in minds*. London: Harper and Row.

Dray, W. 1957: *Laws and explanation in history*. London: Oxford University Press.

Dumont, L. 1970: *Homo hierarchicus*. London: Weidenfeld and Nicholson.

Duncan, J. S. 1978: The social construction of unreality: an interactionist approach to the problem of environment. In D. Ley and M. Samuels (eds.), *Humanistic geography: prospects and problems*. London: Croom Helm, 269–82.

—— 1980: The superorganic in American cultural geography, *Annals of the Association of American Geographers*, 70, 181–98.

—— 1985: Individual action and political power: a structuration perspective. In R. J. Johnston (ed.), *The future of geography*. London: Methuen, 174–89.

—— and Ley, D. 1982: Structural Marxism and human geography: a critical assessment, *Annals of the Association of American Geographers*, 72, 30–59.

Duncan, S. S. and Sayer, R. A. 1980: Debate on geography and the vampire trick, *Area*, 12, 195–7.

Dunford, M. F. 1981: *Historical materialism and geography*. Research Paper in Geography, No. 4. Falmer: University of Sussex.

Eliot Hurst, M. E. 1980: Geography, social science and society: towards a de-definition, *Australian Geographical Studies*, 18, 3–21.

—— 1985: Geography has neither existence nor future. In R. J. Johnston (ed.), *The future of geography*. London: Methuen, 59–91.

Elster, J. 1979: *Ulysses and the Sirens: studies in rationality and irrationality*. Cambridge: Cambridge University Press.

—— 1986: *An introduction to Karl Marx*. Cambridge: Cambridge University Press.

Empson, W. 1953: *Seven types of ambiguity*. London: Chatto and Windus.

Engels, F. 1892: *Socialism, utopian and scientific*. Trans: E. Aveling first pub. 1877; London: Sonnenschien.

Entrikin, N. J. 1976: Contemporary humanism in geography, *Annals of the Association of American Geographers*, 66, 615–32.

—— 1985: Humanism, naturalism and geographical thought, *Geographical Analysis*, 17, 243–7.

Evenson, N. 1973: *Two Brazilian capitals: architecture and urbanism in Rio de Janeiro and Brasilia*. New York: Yale University Press.

Everard, C. E. 1954: The Solent River: a geomorphological study, *Transactions of the Institute of British Geographers*, 20, 41–58.

Everitt, J. C. 1976: Community and propinquity in a city, *Annals of the Association of American Geographers*, 66, 104–16.

Eyles, J. 1979: Social geography and the study of the capitalist city, *Tijdschrift voor Economische en Sociale Geografie*, 59, 296–305.

—— 1981: Why geography cannot be Marxist: towards an understanding of lived experience, *Environment and Planning A*, 1371–88.

—— and Lee, R. 1982: Human geography in explanation, *Transactions of the Institute of British Geographers*, 7, 117–22.

Festinger, L. 1957: *A theory of cognitive dissonance*. Evanston, Ill.: Row, Peterson.

Feyerabend, P. 1975: *Against method: outline of an anarchistic theory of knowledge*. London: New Left Books.

Fine, A. 1984: The natural ontological attitude. In J. Leplin (ed.) *Scientific realism*. Berkeley: University of California Press, 83–107. Reprinted in R. Boyd, P. Gasper, and J. D. Trout (eds.) 1991: *The philosophy of science*, 261–77.

Fish, S. 1979: Normal circumstances, literal language. In P. Rabinow and W. M. Sullivan (eds.), *Interpretive social sciences: a reader*. Berkeley: University of California Press.

Fishbein, M. 1967: Attitude and the prediction of behavior. In M. Fishbein (ed.), *Readings in attitude theory and measurement*. New York: Wiley, 477–92.

—— and Ajzen, I. 1975: *Belief, attitude, intention and behavior*. Reading, Mass.: Addison-Wesley.

Fitzhardinge, L. F. 1954: In search of a capital city. In H. L. White (ed.), *Canberra: a nation's capital*. Canberra: Angus and Robertson, 3–13.

Fleure, H. J. 1947: Some problems of society and environment, *Transactions of the Institute of British Geographers*, 12, 1–37.

Fløistad, G. 1982: Introduction. In G. Fløistad (ed.), *Contemporary philosophy: a new survey*, The Hague: Martinus Nijhoff, ii. 1–17.

Foord, J. and Gregson, N. 1986: Patriarchy: towards a reconceptualisation, *Antipode*, 18, 186–211.

Forster, E. M. 1910: *Howard's End*. Republished 1951; London: Arnold.

—— 1951: *Two cheers for democracy*. London: Arnold.

Foss, M. 1949: *Symbol and metaphor in human experience*. Princeton, NJ: Princeton University Press.

Fotheringhame, A. S. 1981: Spatial structure and distance-decay parameters, *Annals of the Association of American Geographers*, 71, 425–36.

—— 1982: Distance-decay parameters: a reply, *Annals of the Association of American Geographers*, 72, 551–3.

Frazier, J. W. 1981: Pragmatism: geography and the real world. In M. E. Harvey and B. P. Holly (eds.), *Themes in geographic thought*. London: Croom Helm, 61–72.

Frisby, E. M, 1951: Weather–crop relationships: forecasting spring-wheat yield in the northern Great Plains of the United States, *Transactions of the Institute of British Geographers*, 17, 79–96.

Garrison, W. 1960: Connectivity of the interstate highway system, *Papers and Proceedings of the Regional Science Association*, 6, 121–37.

Geography and gender: an introduction to feminist geography: 1984. London: Hutchinson in association with The Explorations in Feminism Collective.

Georgescu-Roegen, N. 1966: *Analytical economics: issues and problems*. Cambridge, Mass.: Harvard University Press.

Geras, N. 1987: Post-Marxism? *New Left Review*, 163, 40–82.

Gerrard, A. J. 1984: Multiple working hypotheses and equifinality in geomorphology: comments on the recent article by Haines-Young and Petch, *Transactions of the Institute of British Geographers*, 9, 364–6.

Gibson, E. M. W. 1981: Realism. In M. E. Harvey and B. P. Holly (eds.), *Themes in geographic thought*. London: Croom Helm, 148–62.

Gibson, J. J. 1979: *The ecological approach to visual perception*. Boston: Houghton-Mifflin.

Giddens, A. 1976: *New rules of sociological method: a positive critique of interpretative sociologies*. London: Hutchinson.

—— 1979: *Central problems in social theory: action, structure and contradiction in social analysis*. London: Macmillan.

—— 1984: *The constitution of society*. Oxford: Polity.

—— 1985: Time, space and regionalization. In D. Gregory and J. Urry (eds.), *Social relations and spatial structures*. London: Macmillan, 265–95.

Gier, J. and Walton, J. 1987: Some problems with reconceptualising patriarchy, *Antipode*, 19, 54–8.

Glacken, C. 1967: *Traces on the Rhodian shore: nature and culture in Western thought from ancient times to the end of the eighteenth century*. Berkeley: University of California Press.

Gold, J. R. 1980: *An introduction to behavioural geography*. Oxford: Oxford University Press.

—— and Goodey, B. 1984: Behavioural and perceptual geography: criticisms and response, *Progress in Human Geography*, 8, 544–50.

Golledge, R. G. 1979: Reality, process and the dialectical relation between man and environment. In S. Gale and G. Olsson (eds.), *Philosophy in geography*. Dordrecht: Reidel, 109–20.

—— 1981: Misconceptions, misinterpretations, and misrepresentations of behavioral approaches in human geography, *Environment and Planning A*, 13, 1325–44.

—— 1985: Teaching behavioural geography, *Journal of Geography in Higher Education*, 9, 111–27.

—— and Stimson, R. J. 1987: *Analytical behavioural geography*. London and Beckenham: Croom Helm.

Gombrich, E. 1960: *Art and illusion*. London: Phaidon.

—— 1979a: *Ideals and idols: essays on values in history and in art*. Oxford: Phaidon.

—— 1979b: *The sense of order: a study in the psychology of decorative art*. Oxford: Phaidon.

Goodey, B. and Gold, J. R. 1985: Behavioural and perceptual geography: from retreospect to prospect, *Progress in Human Geography*, 9, 585–95.

Gorman, R. A. 1977: *The dual vision: Alfred Schutz and the myth of phenomenological social science*. London: Routledge and Kegan Paul.

Goudie, A. 1985: Climatic geomorphology. In A. Goudie, B. W. Atkinson, K. J. Gregory, I. G. Simmons, D. R. Stoddart, and D. Sugden (eds.), *The encyclopaedic dictionary of physical geography*. Oxford: Blackwell, 84.

—— 1986: The integration of human and physical geography, *Transactions of the Institute of British Geographers*, 11, 454–8.

Gould, P. R. 1966: *On mental maps*. Michigan Inter-University Community of Mathematical Geographers Discussion Paper No. 9, Ann Arbor: University of Michigan Press, Department of Geography. Reprinted in R. M. Downs and D. Stea (eds.) , *Image and environment*. Chicago: Aldine, 182–220.

—— 1969: Methodological developments since the fifties, *Progress in Geography*, 1, 1–50.

—— 1981: Social physics. In R. J. Johnston, D. Gregory, P. Haggett, D. Smith, and D. R. Stoddart (eds.), *The dictionary of human geography*. Oxford: Blackwell, 312.

—— and White, R. 1974: *Mental maps*. Harmondsworth: Penguin.

Gould, S. J. 1966: Allometry and size in ontogeny and phylogeny, *Biological Review*, 41, 587–640.

Gouldner, A. W. 1980: *The two Marxisms: contradictions in the development of theory*. London: Macmillan.

Graham, E. 1986: The unity of geography: a comment, *Transactions of the Institute of British Geographers*, 11, 464–7.

—— 1988: *Some implications of realist philosophy for geography*. Paper delivered with D. Livingstone to the Annual Conference, Institute of British Geographers, Loughborough, 7 Jan. (in press).

Graham, J. 1988: Post-modernism and Marxism, *Antipode*, 20, 60–8.

Green, H. L. 1955: Hinterland boundaries of New York city and Boston in southern New England, *Economic Geography*, 31, 283–300.

Gregory, D. 1978: *Ideology, science and human geography*. London: Hutchinson.
—— 1980: The ideology of control: systems theory and geography, *Tijdschrift voor Economische en Sociale Geografie*, 70, 327–42.
—— 1981a: Hermeneutics. In R. J. Johnston, D. Gregory, P. Haggett, D. Smith, and D. R. Stoddart (eds.), *The dictionary of human geography*. Oxford: Blackwell, 145–6.
—— 1981b: Idealism. In R. J. Johnston, D. Gregory, P. Haggett, D. Smith, and D. R. Stoddart (eds.), *The dictionary of human geography*. Oxford: Blackwell, 156–8.
—— 1981c: Human agency and human geography, *Transactions of the Institute of British Geographers*, 6, 1–18.
—— 1981d: Critical theory. In R. J. Johnston, D. Gregory, P. Haggett, D. Smith, and D. R. Stoddart (eds.), *The dictionary of human geography*. Oxford: Blackwell, 59–61.
—— 1986a: Realism. In R. J. Johnston, D. Gregory, and D. M. Smith (eds.), *The dictionary of human geography*. 2nd edn. Oxford: Blackwell, 387–90.
—— 1986b: Structuration theory. In R. J. Johnston, D. Gregory, and D. M. Smith (eds.), *The dictionary of human geography*. 2nd edn. Oxford: Blackwell, 464–9.
—— 1988: The crisis of modernity? Human geography and social theory. Paper delivered to the *Annual Conference, Institute of British Geographers*, Loughborough, 7 Jan.
—— 1992: *The geographical imagination*. In press; Oxford: Blackwell.
—— and Urry, J. 1985: *Social relations and spatial structures*. London: Macmillan.
Gregory, K. J. 1985: *The nature of physical geography*. London: Arnold.
—— 1987: The power of nature: energetics in physical geography. In K. J. Gregory (ed.), *Energetics of physical environment*. Chichester: Wiley, 1–31.
—— and Walling, D. E. (eds.) 1987: *Human activity and environmental processes*. Chichester: Wiley.
Gregory, R. L. 1974: Perceptions as hypotheses. In S. C. Brown (ed.), *Philosophy of psychology*. London: Macmillan, 195–210.
Gregson, N. 1986: On duality and dualism: the case of structuration and time geography, *Progress in Human Geography*, 10, 184–205.
—— 1987: Structuration theory: some thoughts on the possibilities for empirical research, *Environment and Planning D: Society and Space*, 5, 73–91.
—— and Foord, J. 1987: Patriarchy: comments on critics, *Antipode*, 19, 371–5.
Grigg, D. 1965: The logic of regional systems, *Annals of the Association of American Geographers*, 55, 465–91.
—— 1966: Reply comment, *Annals of the Association of American Geographers*, 56, 376–7.
Guelke, L. 1971: Problems of scientific explanation in geography, *The Canadian Geographer*, 15, 38–53.
—— 1974: An idealist alternative in human geography, *Annals of the Association of American Geographers*, 64, 193–202.
—— 1975: On rethinking historical geography, *Area*, 7, 135–8.
—— 1976: The philosophy of idealism, *Annals of the Association of American Geographers*, 66, 168–9.

268 References

—— 1979: Idealist human geography, *Area*, 11, 80–1.
—— 1981: Idealism. In M. E. Harvey and B. P. Holly (eds.), *Themes in geographical thought*. London: Croom Helm, 133–47.
—— 1982: *Historical understanding in geography: an idealist approach*. Cambridge: Cambridge University Press.
Habermas, J. 1971: *Toward a rational society*. London: Heinemann.
Hägerstrand, T. 1970: What about people in regional science? *Papers and Proceedings of the Regional Science Association*, 24, 7–21.
Haggett, P. 1965: *Location analysis in human geography*. London: Arnold.
—— 1978: Spatial forecasting: a view from the touchline. In R. L. Martin, N. J. Thrift, and R. J. Bennett (eds.), *Towards the dynamic analysis of spatial patterns*. London: Pion, 205–10.
—— and Chorley, R. J. 1967: Models, paradigms and the new geography. In R. J. Chorley and P. Haggett (eds.), *Models in geography*. London: Methuen, 19–41.
—— —— 1969: *Network models in geography*. London: Arnold.
—— —— and Stoddart, D. R. 1965: Scale standards in geographical research, a new measure of area magnitude, *Nature*, 205, 844–7.
—— Cliff, A. D., and Frey, A. 1977: *Locational analysis in human geography*, i: *Locational Models*. London: Arnold.
Haines-Young, R. H. and Petch, J. R. 1980: The challenge of critical rationalism for methodology in physical geography, *Progress in Physical Geography*, 4, 63–78.
—— 1983: Multiple working hypotheses: Equifinality and the study of landforms, *Transactions of the Institute of British Geographers*, 8, 458–66.
—— 1984: Multiple working hypotheses: a reply, *Transactions of the Institute of British Geographers*, 9, 367–71.
Hall, A. D. and Fagen, R. E. 1956: Definition of system, *General Systems*, 1, 18–29.
Hall, E. T. 1966: *The hidden dimension*. New York: Doubleday.
Hamilton, F. E. I. 1974: A view of spatial behaviour, industrial organizations and decision-making. In F. E. I. Hamilton (ed.), *Spatial perspectives on industrial organization and decision-making*. London: Wiley, 3–43.
Hammett, D. 1974: *The Maltese falcon*. London: Cassell.
Harris, C. 1971: Theory and synthesis in historical geography, *The Canadian Geographer*, 15, 157–72.
—— 1978: The historical mind and the practice of geography. In D. Ley and M. Samuels (eds.), *Humanistic geography: prospects and problems*. London: Croom Helm, 123–37.
Harrison, R. T. and Livingstone, D. N. 1979: There and back again: towards a critique of idealist human geography, *Area*, 11, 75–9 and 81–2.
—— 1980: Philosophy and problems in human geography: a presuppositional approach, *Area*, 12, 25–31.
Hart, C. 1984: Geographical education: does it help to reason? *New Zealand Journal of Geography*, 77, 14–17.
Hartmann, H. 1981: The unhappy marriage of marxism and feminism: towards a more progressive union. In L. Sargent (ed.), *Women and Revolution*. London: Pluto Press.

Hartshorne, R. 1939: *The nature of geography*. Lancaster, Penn.: Association of American Geographers.
—— 1954: Comment on 'Exceptionalism in geography', *Annals of the Association of American Geographers*, 44, 108–9.
—— 1955: 'Exceptionalism in geography' re-examined, *Annals of the Association of American Geographers*, 45, 205–44.
—— 1959: *Perspective on the nature of geography*. Chicago: Rand McNally.
Harvey, D.W. 1967: Models of the evolution of spatial patterns in geography. In R.J. Chorley and P. Haggett (eds.), *Models in geography*. London: Methuen, 549–608.
—— 1968: Pattern, process and the scale problem in geographical research, *Transactions of the Institute of British Geographers*, 45, 71–8.
—— 1969a: *Explanation in geography*. London: Arnold.
—— 1969b: Conceptual and measurement problems in the cognitive-behavioral approach to location theory. In K.R. Cox, and R.G. Golledge (eds.), *Behavioral problems in geography*. Studies in Geography, No. 17, Evanston: Northwestern University, Department of Geography, 35–68.
—— 1969c: Review of A. Pred (1967), *Behavior and location: foundations for a geographic and dynamic location theory, part I*, *Geographical Review*, 59, 312–14.
—— 1971: Social processes, spatial form, and the redistribution of real income in an urban system. In M. Chisholm, A.E. Frey, and P. Haggett (eds.), *Regional forecasting: Proceedings of the twenty-second symposium of the Colston Research Society*. London: Butterworth, 267–300.
—— 1972a: Revolutionary and counter-revolutionary theory in geography and the problem of ghetto formation, *Antipode*, 4, 1–13 and 36–41.
—— 1972b: On obfuscation in geography: a comment on Gale's heterodoxy, *Geographical Analysis*, 4, 323–30.
—— 1972c: Review of P. Wheatley, *The pivot of the four quarters: a preliminary enquiry into the origins and character of the ancient Chinese city*, *Annals of the Association of American Geographers*, 62, 509–13.
—— 1973: *Social justice and the city*. London: Arnold.
—— 1977: Communication on recent comments by Professor Carter, *The Professional Geographer*, 29, 405–7.
—— 1979: Monument and myth, *Annals of the Association of American Geographers*, 69, 362–81. Reprinted in Harvey, 1985.
—— 1981: Marxist geography. In R.J. Johnston, D. Gregory, P. Haggett, D. Smith, and D.R. Stoddart (eds.), *The dictionary of human geography*. Oxford: Blackwell, 209–12.
—— 1983: Geography. In T. Bottomore (ed.), *A dictionary of Marxist thought*. Oxford: Blackwell, 189–92.
—— 1984: On the history and present condition of geography: an historical materialist manifesto, *The Professional Geographer*, 36, 1–11.
—— 1985: *Consciousness and the urban experience*. Oxford: Blackwell.
—— 1987: Three myths in search of a reality in urban studies, *Environment and Planning D: Society and Space*, 5, 367–76.
—— 1988: From models to Marx: notes on the project to 're-model' contemporary geography. In W.D. MacMillan (ed.), *Re-modelling geography*. In press; Oxford: Blackwell.

270 References

Hay, A. 1979: Positivism in human geography: response to critics. In D. T. Herbert and R. J. Johnston (eds.), *Geography and the urban environment: progress in research and applications*, London: Wiley, ii. 1–26.

Heathcote, R. L. 1987: Images of a desert? Perceptions of arid Australia, *Australian Geographical Studies*, 25, 3–25.

Heidegger, M. 1927: *Being and time*. Trans. J. Macquarrie and E. Robinson 1962. Oxford: Blackwell.

Helson, H. 1964: *Adaptation level theory: an experimental and systematic approach to behavior*. New York: Harper and Row.

Hempel, C. G. 1942: The function of general laws in history, *Journal of Philosophy*, 39, 35–48.

Hepple, L. W. 1981a: Entropy-maximizing models. In R. J. Johnston, D. Gregory, P. Haggett, D. M. Smith, and D. R. Stoddart (eds.), *The dictionary of human geography*. Oxford: Blackwell, 102–3.

—— 1981b: Lowry model. In R. J. Johnston, D. Gregory, P. Haggett, D. M. Smith, and D. R. Stoddart (eds.), *The dictionary of human geography*. Oxford: Blackwell, 196–7.

Hewitt, K. 1983: The idea of calamity in a technocratic age. In K. Hewitt (ed.), *Interpretations of calamity*. Boston: Allen and Unwin, 4–32.

Hildreth, C. 1967: Review of N. Georgescu-Roegen, *Analytical economics, American Economic Review*, 57, 589–95.

Hiley, D. R., Bohman, J. F., and Shusterman, R. (eds.) 1991: *The interpretive turn: philosophy, science, culture*. Ithaca: Cornell University Press.

Hill, M. R. 1981: Positivism: a 'hidden' philosophy in geography. In M. E. Harvey and B. P. Holly (eds.), *Themes in geographic thought*. London: Croom Helm, 38–60.

Hodder, B. W. 1965: Some comments on the origins of traditional markets south of the Sahara, *Transactions of the Institute of British Geographers*, 36, 97–105.

Hooson, D. J. M. 1984: The Soviet Union. In R. J. Johnston and P. Claval (eds.), *Geography since the Second World War: an international survey*. London: Croom Helm, 79–106.

Hoselitz, B. F. 1960: Theories of stage of economic growth. In B. F. Hoselitz, J. J. Spengler, J. M. Letiche, E. McKinley, J. Buttrick, and H. J. Bruton (eds.), *Theories of economic growth*. Glencoe, Ill.: Free Press, 193–238.

Hudson, L. 1966: *Contrary imaginations*. London: Methuen.

Hudson, R. 1987: Flying the flag for flagging geography, *The Times Higher Educational Supplement*, 23 Oct., 18.

Hufferd, J. 1980: Idealism and the participant's world, *The Professional Geographer*, 32, 1–5.

Huggett, R. 1980: *Systems analysis in geography*. Oxford: Clarendon Press.

—— 1981: A hard line on soft systems, *Area*, 13, 224–7.

Ihde, D. 1977: *Experimental phenomenology: an introduction*. New York: Putnam.

Ilyichev, L. F. 1964: Remarks about a unified geography, *Soviet Geography*, 5 Apr., 32–46.

Jackson, P. 1981: Phenomenology and social geography, *Area*, 13, 299–305.

—— and Smith, S. J. 1984: *Exploring social geography*. London: Allen and Unwin.

Jackson, R. T. 1971: Periodic markets in southern Ethiopia, *Transactions of the Institute of British Geographers*, 53, 31–42.

James, W. 1920: *Collected essays and reviews*. First pub. 1878; London: Longman.

Janelle, D. G. 1968: Central-place development in a time–space framework, *The Professional Geographer*, 20, 5–10.

—— 1969: Spatial reorganization: a model and concept, *Annals of the Association of American Geographers*, 59, 348–64.

Jantsch, E. 1980: *The self-organizing universe*. Oxford: Pergamon.

Jaspers, K. 1969: *Philosophy*, i. Chicago: University of Chicago Press.

Jennings, J. N. 1952: *The origin of the Broads*. Royal Geographical Society Research Series, No. 2. London: Royal Geographical Society.

—— and Lambert, J. M. 1953: The origin of the Broads, *Geographical Journal*, 119, 91.

Jensen-Butler, C. 1972: Gravity models as planning tools: a review of theoretical and operational problems, *Geografiska Annaler*, 54B, 68–78.

—— 1981: *A critique of behavioural geography: an epistemological analysis of cognitive mapping and of Hägerstrand's time–space model*. Aarhus: Aarhus University, Geographical Institute.

Johnson, H. G. 1971: The Keynesian revolution and the monetarist counter-revolution, *American Economic Review*, 61, (Papers and Proceedings), 1–14.

Johnson, L. 1987: (Un)realist perspectives: patriarchy and feminist challenges in geography, *Antipode*, 19, 210–15.

Johnston, R. J. 1976: *Classification in geography*. Concepts and Techniques in Modern Geography (Catmog), 6. Norwich: Geo Books.

—— 1978: Paradigms and revolutions or evolution: observations on human geography since the Second World War, *Progress in Human Geography*, 2, 189–206.

—— 1979a: *Political, electoral, and spatial systems*. Oxford: Clarendon.

—— 1979b, 1983d (2nd edn.), 1987 (3rd edn.): *Geography and geographers: Anglo-American human geography since 1945*. London: Arnold.

—— 1980: On the nature of explanation in human geography, *Transactions of the Institute of British Geographers*, 5, 402–12.

—— 1981: Ecological fallacy. In R. J. Johnston, D. Gregory, P. Haggett, D. Smith and D. R. Stoddart (eds.), *The dictionary of human geography*, Oxford: Blackwell, 89.

—— 1983a: On geography and the history of geography. *History of Geography Newsletter*, 3, 1–7. Washington: Association of American Geographers.

—— 1983b: *Philosophy and human geography: an introduction to contemporary approaches*. London: Arnold.

—— 1983c: Resource analysis, resource management and the integration of human and physical geography, *Progress in Physical Geography*, 7, 127–46.

—— 1986a: Four fixations and the quest for unity in geography, *Transactions of the Institute of British Geographers*, 11, 449–53.

—— 1986b: Review of J. L. Paterson, David Harvey's geography, *Antipode*, 18, 96–109.

—— (ed.) 1985: *The future of geography*. London: Methuen.

—— and Claval, P. (eds.), 1984: *Geography since the Second World War: an international survey*. London: Croom Helm.

—— Gregory, D., Haggett, P., Smith, D., and Stoddart, D.R. (eds.) 1981: *The dictionary of human geography*. Oxford: Blackwell.

Jonas, A. 1988: A new regional geography of localities, *Area*, 20, 101–10.

Jones, D.K.C. 1980: The Tertiary evolution of south-east England with particular reference to the Weald. In D.K.C. Jones (ed.), *The shaping of southern England*. London: Academic Press, 13–47.

Jones, E. 1956: Cause and effect in human geography, *Annals of the Association of American Geographers*, 46, 369–77.

Jones, K. and Moon, G. 1987: *Health, disease and society: an introduction to medical geography*. London: Routledge and Kegan Paul.

Jones, T.P. and McEvoy, D. 1978: Race and space in cloud-cuckoo land, *Area*, 10, 162–6.

Joynt, C.B. and Rescher, N. 1961: The problem of uniqueness in history, *History and Theory*, 1, 150–62.

Kansky, K.J. 1963: *Structure of transportation networks*. Department of Geography Research Paper, No. 84. Chicago: University of Chicago.

Kant, I. 1781: *Critique of pure reason*. Trans. N.K. Smith 1978. London: Macmillan.

Kates, R.W. 1962: *Hazard and choice perception in flood plain management*. Department of Geography Research Paper, No. 78. Chicago: University of Chicago.

Kauth, R.J. and Thomas, G.S. 1976: The Tasseled Cap: a graphic description of the spectral-temporal development of agricultural crops as seen by Landsat, *Proceedings of the Symposium on Machine Processing of Remotely Sensed Data*. West Lafayette, Ind.: Purdue University, 4B41–4B51.

Keat, R. 1981: *The politics of social theory.*. Oxford: Blackwell.

—— and Urry, J. 1982: *Social theory as science*. London: Routledge and Kegan Paul.

Kellerman, A. 1987: Structuration theory and attempts at integration in human geography, *The Professional Geographer*, 39, 267–74.

Kennedy, B.A. 1977: A question of scale, *Progress in Physical Geography*, 1, 154–7.

—— 1979: A naughty world, *Transactions of the Institute of British Geographers*, 4, 550–8.

Khosla, R. 1971: Chandigarh: dream and reality, *Geographical Magazine*, 43, 679–83.

King, R. 1982: Marxist and radical geographical literature in English: an introduction. In M. Quaini (ed.), *Geography and Marxism*. Oxford: Blackwell, 175–200.

Kirby, A.M. and Pinch, S.P. 1983: Territorial justice and service allocation. In M. Pacione (ed.), *Progress in urban geography*. London: Croom Helm, 223–50.

Kirk, W. 1951: Historical geography and the concept of the behavioural environment, *Indian Geographical Journal*, Silver Jubilee volume, 152–60.

—— 1963: Problems of geography, *Geography*, 48, 357–71.

—— 1978: The road from Mandalay: towards a geographical philosophy, *Transactions of the Institute of British Geographers*, 3, 381–94.

Klaassen, L.H., Paelinck, J.H.P., and Wagenaar, S. 1979: *Spatial systems: a general introduction*. Farnborough: Saxon House.

Knopp, L. and Lauria, M. 1987: Gender relations as a particular form of social relations, *Antipode*, 19, 48–53.

Koestler, A. 1964: *The act of creation*. London: Hutchinson.

— and Smythies, J. R. (eds.) 1969: *Beyond reductionism*. New York: Macmillan.

Krumme, G. 1969: Toward a geography of enterprise, *Economic Geography*, 45, 30–40.

Kuhn, T. S. 1970: *The structure of scientific revolutions*. First pub. 1962; Chicago: University of Chicago Press.

Laclau, E. and Mouffe, C. 1985: *Hegemony and socialist strategy: towards a radical democratic politics*. London: Verso.

— 1987: Post-Marxism without apologies, *New Left Review*, 166, 79–106.

Lakatos, I. 1970: Falsification and the methodology of scientific research programmes. In I. Lakatos and A. Musgrave (eds.), *Criticism and the growth of knowledge*. Cambridge: Cambridge University Press, 91–195.

— 1978: *The methodology of scientific research programmes*. Cambridge: Cambridge University Press.

— and Musgrave, A. (eds.) 1970: *Criticism and the growth of knowledge*. Cambridge: Cambridge University Press.

Lambert, J. M. and Jennings, J. N. 1960: *The making of the Broads: a reconsideration of their origin in the light of new evidence*. Royal Geographical Society Research Series, No. 3. London: Murray.

Langton, J. 1972: Potentialities and problems of adopting a systems approach to the study of change in human geography, *Progress in Human Geography*, 4, 125–79.

Laudan, L., 1977: *Progress and its problems: towards a theory of scientific growth*. Berkeley: University of California Press.

Lavrov, S. B., Preobrazhenskiy, V. S., and Sdasyak, G. V. 1980: Radical geography: its roots, history and positions, *Soviet Geography*, 21, 308–21, Trans.; first pub. 1979.

Leach, E. R. 1976: *Culture and communication: the logic by which symbols are connected*. Cambridge: Cambridge University Press.

Lee, T. R. 1968: Urban neighbourhood as a socio-spatial schema, *Human Relations*, 21, 241–67.

— 1978: Race, space and scale, *Area*, 10, 365–7.

Lefebvre, H. 1972: *La pensée marxiste et la ville*. Tournai: Casterman.

Leopold, L. B. 1960: Ecological systems and water resources. *Proceedings of the sixth Biennial Wilderness Conference*, Sierra Club, San Francisco.

Levison, M., Fenner, T. I., Sentance, W. A., Ward, R. G., and Webb, J. W. 1969: A model of accidental drift voyaging in the Pacific Ocean with applications to the Polynesian colonization problem, *Proceedings of the International Federation for Information Processing Congress*, 1968, 1521–6.

Lévi-Strauss, C. 1969: *The elementary structures of kinship*. London: Eyre and Spottiswoode.

Ley, D. 1977: Social geography and the taken-for-granted world, *Transactions of the Institute of British Geographers*, 2, 498–512.

— 1978: Social geography and social action. In D. Ley and M. Samuels (eds.), *Humanistic geography: prospects and problems*. London: Croom Helm, 41–57.

274 References

—— and Samuels, M.S. (eds.) 1978: *Humanistic geography: prospects and problems*. London: Croom Helm.

Livingstone, D.N., and Graham, E. 1988: Some implications of realist philosophy for geography. Paper delivered to the *Annual Conference, Institute of British Geographers*, Loughborough, 7 Jan. (in press).

—— and Harrison, R.T. 1980: The frontier: metaphor, myth, and model, *The Professional Geographer*, 32, 127–32.

Lloyd, R. 1982: A look at images, *Annals of the Association of American Geographers*, 72, 532–48.

Lombardo, T.J. 1987: *The reciprocity of perceiver and environment: the evolution of James J. Gibson's ecological psychology*. Hillside, NJ: Erlbaum.

Lowenthal, D. 1961: Geography, experience, and imagination: towards a geographical epistemology, *Annals of the Association of American Geographers*, 51, 241–60.

—— 1965: Introduction to G.P. Marsh, *Man and nature*. Cambridge, Mass.: Harvard University Press, pp. ix–xxix.

Lowry, I.S. 1964: *A model of metropolis*. RX-4035L. Santa Monica: Rand Corporation.

Lukermann, F. 1958: Towards a more geographic economic geography. *The Professional Geographer*, 10, 2–10.

Lynch, K. 1960: *The image of the city*. Cambridge, Mass.: MIT Press.

Lyotard, J.F. 1987: The postmodern condition. In K. Baynes, J. Bohman, and T. McCarthy (eds.), *After philosophy: end or transformation?* Cambridge, Mass.: MIT Press 73–94.

McDowell, L. 1986: Beyond patriarchy: a class-based explanation of women's subordination, *Antipode*, 18, 311–21.

MacLachlan, I. 1981: Settlement pattern evolution and Catastrophe Theory: a comment, *Transactions of the Institute of British Geograhers*, 6, 126–8.

McNee, R.B. 1974: A systems approach of understanding the geographic behaviour of organizations, especially large corporations. In F.E.I. Hamilton (ed.), *Spatial perspectives on industrial organization and decision-making*. London: Wiley, 47–75.

McQuaid, C. 1987: Geography maps a dismal future, *The Times Higher Educational Supplement*, 2 Oct., 9.

Mair, A. 1986: Thomas Kuhn and understanding geography, *Progress in Human Geography*, 10, 344–69.

Makkreel, R.A. 1975: *Dilthey: philosopher of human studies*. Princeton, NJ: Princeton University Press.

Malachowski, A. 1990: Deep epistemology without foundations (in language). In Malachowski, A. (ed.), *Reading Rorty*, 139–55.

—— (ed.) 1990: *Reading Rorty: critical responses to* Philosophy and the mirror of nature *(and beyond)*. Oxford: Blackwell.

Manicas, P.T. 1987: *A history and philosophy of the social sciences*. Oxford: Blackwell.

Marchand, B. 1979: Dialectics and geography. In S. Gale and G. Olsson (eds.), *Philosophy in geography*. Dordrecht: Reidel, 237–67.

Marcus, S. 1974: Introduction to D. Hammett, *The continental op*. New York: Random House.

Marcuse, H. 1972: *One dimensional man*. London: Abacus.

Marsh, G. P. 1965: *Man and nature*. First pub. 1864; Cambridge, Mass.: Harvard University Press.

Martin A. F. 1951: The necessity for determinism: a metaphysical problem confronting geographers, *Transactions of the Institute of British Geographers*, 17, 1–11.

Martin, B. 1981: The scientific straitjacket: the power structure of science and the suppression of environmental scholarship, *The Ecologist*, 11, 33–43.

Maruyama, M. 1963: The second cybernetics: deviation amplifying mutual causal processes, *American Scientist*, 51, 164–79.

—— 1973: A new logical model for futures research, *Futures*, 5, 435–7.

Marx, K. 1962: Introduction to *A contribution to the critique of political economy*, in *Karl Marx and Frederick Engels: selected works*, 2 vols. First pub. 1859; Moscow: Foreign Languages Publishing House.

Maslow, A. H. 1967: The creative attitude. In R. L. Mooney and T. A. Razik (eds.), *Explorations in creativity*. New York: Harper and Row, 43–54.

Mason, H. 1991: Review [of the first edition of this book], *Geography*, 76, 89.

Massey, D. 1975: Behavioural research, *Area*, 7, 201–3.

—— 1984: *Spatial divisions of labour: social structures and the geography of production*. London: Macmillan.

—— 1987: Why geography matters, *Geography Review*, 1, 2–7.

—— and Meegan, R. A. 1979: The geography of industrial reorganization, *Progress in Planning*, 10, 155–237.

—— 1985 Introduction: the debate. In D. Massey and R. Meegan (eds.), *Politics and method: contrasting studies in industrial geography*. London: Methuen, 1–12.

Masterman, M. 1970: The nature of a paradigm. In I. Lakatos and A. Musgrave (eds.), *Criticism and the growth of knowledge*. Cambridge: Cambridge University Press, 59–90.

Matley, I. M. 1966: The Marxist approach to the geographical environment, *Annals of the Association of American Geographers*, 56, 97–111.

Matthews, J. A. 1985a: Polyclimax. In A. Goudie, B. W. Atkinson, K. J. Gregory, I. G. Simmons, D. R. Stoddart, and D. Sugden (eds.), *The encyclopaedic dictionary of physical geography*. Oxford: Blackwell, 341.

—— 1985b: Climax vegetation. In A. Goudie, B. W. Alkinson, K. J. Gregory, I. G. Simmons, D. R. Stoddart, and D. Sugden (eds.), *The encyclopaedic dictionary of physical geography*. Oxford: Blackwell, 85–6.

May, J. A. 1982: On orientations and reorientations in the history of western geography. In J. D. Wood (ed.), *Rethinking geographical inquiry*. Geographical Monographs, No. 11. York: York University, 31–72.

Mazey, M. E. and Lee, D. R. 1983: *Her space, her place: a geography of women*. Resource Publications in Geography. Washington: Association of American Geographers.

Medawar, P. B. 1967: *The art of the soluble*. London: Methuen.

—— 1969: *Induction and intuition in scientific thought*. London: Methuen.

—— 1979: *Advice to a young scientist*. London: Harper and Row.

Meddin, J. 1975: Attitudes, values and related concepts: a system of classification, *Social Science Quarterly*, 55, 889–900.

Mercer, D.C. 1984: Unmasking technocratic geography. In M. Billinge, D. Gregory, and R. Martin (eds.), *Recollections of a revolution: geography as a spatial science.* London: Macmillan, 153–99.

—— 1985: *Reading the book of nature: physical and human geography and the limits of science.* Working Paper, No. 18. Melbourne: Monash University, Department of Geography.

—— and Powell, J.M. 1972: Phenomenology and related non-positivistic viewpoints in the social sciences. *Monash Publications in Geography*, 1. Clayton, Victoria: Monash University, Department of Geography.

Mesarovic, M.D. (ed.) 1964: *Views on general systems theory.* New York: Wiley.

Miehle, W. 1958: Link-length minimization in networks, *Journal of the Operations Research Society of America*, 6, 232–43.

Miller, G., Gallanter, E., and Pribham, K. 1960: *Plans and the structure of behavior.* New York: Holt, Rinehart, and Wilson.

Miller, J.G. 1965: Living systems: basic concepts, *Behavioral Science*, 10, 193–237.

Mills, C.W. 1959: *The sociological imagination.* New York: Oxford University Press.

Mills, W.J. 1982: Positivism reversed: the relevance of Giambattista Vico, *Transactions of the Institute of British Geographers*, 7, 1–14.

Miner, H. 1952: The folk-urban continuum, *American Sociological Review*, 17, 529–37.

Mitchell, J.C. 1983: Case and situation analysis, *The Sociological Review*, 31, 187–211.

Moles, A.A. 1957: *La création scientifique.* Geneva: Kister.

—— 1964: Le contenu d'une méthodologie appliquée: un essai de recensement des méthodes. In R. Caude and A. Moles (eds.), *Méthodologie vers une science de l'action.* Paris: Gauthier-Villars, 45–89.

—— and Rohmer, E. 1972: *Psychologie de l'espace.* Tournai: Casterman.

Montaigne, M. de 1926: Apologie de Raimond Sebond, in *Œuvres complètes*, iv. First pub. 1588 and 1596, Paris: Conard, 1–122.

Mooij, J.J.A. 1976: *A study of metaphor: on the nature of metaphorical expressions with special reference to their reference.* Amsterdam: North Holland.

Moore, G.T. and Golledge, R.G. 1976: Environmental knowing: concepts and theories. In G.T. Moore and R.G. Golledge (eds.), *Environmental knowing.* Stroudsburg: Dowden, Hutchinson, and Ross, 3–24.

Moos, A.I. and Dear, M.J. 1986: Structuration theory in urban analysis; i: Theoretical exegesis, *Environment and Planning A*, 18, 231–52.

Morgan, R.K. 1981: Systems analysis: a problem of methodology? *Area*, 13, 219–23 and 227–30.

Moss, R.P. 1977: Deductive strategies in geographical generalization, *Progress in Physical Geography*, 1, 23–39.

Murphy, R.L. and Vance, J.E. 1954: Delimiting the CBD, *Economic Geography*, 30, 301–36.

Neisser, U. 1976: *Cognition and reality.* San Francisco: Freeman.

North, D.C. 1955: Location theory and regional economic growth, *Journal of Political Economy*, 63, 243–58.

Norton, W. 1984: *Historical analysis in geography*. London: Longman.

Nowell-Smith, P.H. 1981: History as patterns of thought and action. In L. Pompa and W.H. Dray (eds.), *Substance and form in history: a collection of essays on philosophy of history*. Edinburgh: Edinburgh University Press, 145–55.

Nystuen, J.D. 1963: Identification of some fundamental spatial concepts, *Papers of the Michigan Academy of Science, Arts and Letters*, 48, 373–83.

Ogden, C.K. 1967: *Opposition: a linguistic and psychological analysis*. First pub. 1932; Bloomington: Indiana University Press.

O'Hear, A. 1980: *Karl Popper*. London. Routledge and Kegan Paul.

Ollman, B. 1971: *Alienation: Marx's conception of man in capitalist society*. Cambridge: Cambridge University Press.

Olsson, G. 1975: *Birds in egg*. University of Michigan Department of Geography Publication No. 15. Ann Arbor, Mich.: University of Michigan.

—— 1978: Of ambiguity or far cries from a memorializing mamafesta. In D. Ley and M. Samuels (eds.), *Humanistic geography: prospects and problems*. London: Croom Helm, 109–20.

—— 1979: Social science and human action or on hitting your head against the ceiling of language. In S. Gale and G. Olsson (eds.), *Philosophy in geography*. Dordrecht: Reidel, 287–307.

—— 1980: *Birds in egg: eggs in bird*. London: Pion.

—— 1982: Languages and dialectics. In P.R. Gould and G. Olsson (eds.), *A search for common ground*. London: Pion, 222–31.

Orme, A. 1985: Understanding and predicting the physical world. In R.J. Johnston (ed.), *The future of geography*. London: Methuen, 258–75.

Owens, P.L. and House, R. 1984: *Radical geography: an annotated bibliography*. Norwich: Geo Books.

Pahl, R.E. 1968: The rural–urban continuum. In R.E. Pahl (ed.), *Readings in urban sociology*. London: Pergamon, 263–97.

Paivio, A. 1969: Mental imagery in associative learning and memory, *Psychological Review*, 76, 241–63.

—— 1971: *Imagery and verbal processes*. New York: Holt, Rinehart, and Wilson.

Park, D.A. 1980: *The image of eternity: roots of time in the physical world*. Amherst, Mass.: University of Massachusetts Press.

Paterson, J.H. 1970: Muddy pool or crystal fountain: a note on the sources of our present thought, *St Andrews Geographer*, 1, 6–9.

—— 1975: Other laws, other landscapes. An inaugural lecture delivered in the University of Leicester, 11 Nov. Leicester: University of Leicester, Department of Geography; mimeograph.

Paterson, J.L. 1984: *David Harvey's geography*. London: Croom Helm.

—— 1986: Reply to R.J. Johnston, *Antipode*, 18, 109–15.

Pattee, H.H. 1978: The complementarity principle in biological and social structures, *Journal of Social and Biological Structures*, 1, 191–200.

Peach, C. and Smith, S.J. 1981: Introduction. In C. Peach, V. Robinson, and S.J. Smith (eds.), *Ethnic segregation in cities*. London: Croom Helm, 9–24.

Peake, L. and Jackson, P. 1988: 'The restless analyst': an interview with David Harvey, *Journal of Geography in Higher Education*, 12, 5–20.

278 References

Peet, R. 1977: Introduction. In R. Peet (ed.), *Radical geography: alternative viewpoints on contemporary social issues*. London: Methuen, 1–4.
— 1978: The dialectics of radical geography: a reply to Gordon Clark and Michael Dear, *The Professional Geographer*, 30, 360–4.
Peirce, C. S. 1931–5: *The collected papers of Charles Sanders Peirce*, i–vi, ed. C. Hartshorne and P. Weiss; vii–viii, ed. A. W. Burks. Cambridge, Mass.: Harvard University Press.
Peltier, L. C. 1950: The geographic cycle in periglacial regions as it is related to climatic geomorphology, *Annals of the Association of American Geographers*, 40, 214–36.
Penning-Rowsell, E. C. and Townshend, J. R. G. 1978: The influence of scale on the factors affecting stream channel slope, *Transactions of the Institute of British Geographers*, 3, 395–415.
Pepper, D. 1984: *The roots of modern environmentalism*. London: Croom Helm.
Philbrick, A. K. 1957: Principles of areal functional organisation in regional human geography, *Economic Geography*, 33, 299–366.
Pickles, J. 1985: *Phenomenology, science and geography: spatiality and the human sciences*. Cambridge: Cambridge University Press.
— 1986: *Geography and humanism*. Concepts and Techniques in Modern Geography (Catmog), No. 44. Norwich: Geo Books.
Pipkin, J. S. 1981: Cognitive behavioral geography and repetitive travel. In K. R. Cox and R. G. Golledge (eds.), *Behavioral problems in geography revisited*. London: Methuen, 145–81.
Platt, R. S. 1948: Environmentalism versus geography, *American Journal of Sociology*, 53, 351–8.
Pocock, D. C. D. (ed.) 1981: *Humanistic geography and literature: essays on the experience of place*. London: Croom Helm.
— 1983: The paradox of humanistic geography, *Area*, 15, 355–8.
Popper, K. R. 1959: *The logic of scientific discovery*. First pub. 1934 as *Logik der Forschung*. London: Hutchinson.
— 1961: *The poverty of historicism*. London: Routledge and Kegan Paul.
— 1962: *The open society and its enemies*. 2 vols. London: Routledge and Kegan Paul.
— 1963: *Conjectures and refutations*. London: Routledge and Kegan Paul.
— 1970: Normal science and its dangers. In I. Lakatos, and A. Musgrave (eds.), *Criticisms and the growth of knowledge*. Cambridge: Cambridge University Press, 51–8.
— 1972: *Objective knowledge*. Oxford: Oxford University Press.
— 1974: Replies to my critics. In P. A. Schilpp (ed.), *The philosophy of Karl Popper*, i and ii. Library of Living Philosophers; La Salle, Ind.: Open Court, 961–1197. (The 'intellectual biography' of this published separately, see Popper, 1976.)
— 1976a: *Unended quest*. London: Fontana.
— 1976b: The death of theories and of ideologies. Paper read in Sparta, 14 April. Mimeographed.
— 1986: *Ideologies are a real threat to physical science*. Medawar Lecture given at the Royal Society, London, 12 June.
— and Eccles, J. C. 1977: *The self and its brain*. Berlin: Springer.

Porteous, J.D. 1976: Home: the territorial core, *Geographical Review*, 66, 383–90.
— 1985: Literature and humanist geography, *Area*, 17, 117–22.
Pred, A. 1977: The choreography of existence: comments on Hägerstrand's time-geography, *Economic Geography*, 53, 207–21.
— 1982: Social reproduction and the time-geography of everyday life. In P. Gould and G. Olsson (eds.), *A search for common ground*. London: Pion, 157–86.
— 1984: Place as historically contingent process: structuration and the time-geography of becoming places, *Annals of the Association of American Geographers*, 74, 279–97.
Price, E.T. 1968: The central courthouse square in the American county seat, *Geographical Review*, 58, 29–60.
Prince, H.C. 1971: Real, imagined and abstract worlds of the past, *Progress in Geography*, 3, 1–86.
Quaini, M. 1982: *Geography and Marxism*. Oxford: Blackwell.
Rapoport, A. and Kantor, R.E. 1967: Complexity and ambiguity in environmental design, *Journal of the American Institute of Planners*, 33, 210–21.
Reed, E.S. 1987: Why do things look as they do? The implications of J.J. Gibson's *The ecological approach to visual perception*. In A. Costall and A. Still (eds.), *Cognitive psychology in question*. Brighton: Harvester.
Rees, R. 1973: Geography and landscape painting: an introduction to a neglected field, *Scottish Geographical Magazine*, 89, 147–57.
— 1976a: Images of the prairie: landscape painting and perception in the western interior of Canada, *The Canadian Geographer*, 20, 259–78.
— 1976b John Constable and the art of geography, *Geographical Review*, 16, 59–72.
Reif, B. 1973: *Models in urban and region planning*. Aylesbury: Hill.
Reilly, W.J. 1931: *The law of retail gravitation*. New York: Knickerbocker.
Relph, E.C. 1970: An inquiry into the relations between phenomenology and geography, *The Canadian Geographer*, 14, 193–201.
— 1976: *Place and placelessness*. London: Pion.
— 1981a: Phenomenology. In M.E. Harvey and B.P. Holly (eds.), *Themes in geographic thought*. London: Croom Helm, 99–114.
— 1981b: *Rational landscapes and humanistic geography*. London: Croom Helm.
— 1987: *The modern urban landscape*. London: Croom Helm.
Rescher, N. 1977: *Methodological pragmatism: a systems theoretic approach to the theory of knowledge*. Oxford: Blackwell.
— 1978: *Scientific progress*. Pittsburgh: University of Pittsburgh Press.
— 1979a: Reply to Barker. In E. Sosa (ed.), *The philosophy of Nicholas Rescher: discussion and replies*. Dordrecht: Reidel, 17–18.
— 1979b: Reply to Bonjour. In E. Sosa (ed.), *The philosophy of Nicholas Rescher: discussion and replies*. Dordrecht: Reidel, 173–4.
— 1984: *The limits of science*. Berkeley: University of California Press.
Richardson, M. 1981: Commentary on 'The superorganic in American cultural geography', *Annals of the Association of American Geographers*, 71, 284–91.
Rieser, R. 1973: The territorial illusion and the behavioural sink: critical notes on behavioural geography, *Antipode*, 5, 52–7.

Rorty, R. 1979: *Philosophy and the mirror of nature*. Princeton: Princeton University Press.

— 1986: The contingency of language, *London Review of Books*, 8, 17 April.

Rose, A.J. 1966: Dissent from down under: metropolitan primacy as the normal state, *Pacific Viewpoint*, 7, 1–27.

Rose, C. 1981: William Dilthey's philosophy of historical understanding: a neglected heritage of contemporary humanistic geography. In D.R. Stoddart (ed.), *Geography, ideology and social concern*. Oxford: Blackwell, 99–133.

Rostow, W.W. 1960: *The stages of economic growth: a non-Communist manifesto*. Cambridge: Cambridge University Press.

Rowles, G.D. 1978a: *The prisoners of space? Exploring the geographical experience of older people*. Boulder, Colo.: Westview.

— 1978b: Reflections on experimental field work. In D. Ley and M. Samuels (eds.), *Humanistic geography: prospects and problems*. London: Croom Helm, 173–93.

Rubinoff, L. 1970: *Collingwood and the reform of metaphysics: a study in the philosophy of mind*. Toronto: University of Toronto Press.

Russell, B. 1923: On vagueness, *Australasian Journal of Psychology and Philosophy*, 1, 84–92.

Sack, R.D. 1973: A concept of physical space in geography, *Geographical Analysis*, 5, 16–34.

— 1978: Geographic and other views of space. In K.W. Butzer (ed.), *Dimensions of human geography*. Department of Geography Research Paper No. 186. Chicago: University of Chicago, 166–84.

Samuels, M.S. 1978: Existentialism and human geography. In D. Ley and M. Samuels (eds.), *Humanistic geography: prospects and problems*. London: Croom Helm, 22–40.

— 1981: An existential geography. In M.E. Harvey and B.P. Holly (eds.), *Themes in geographic thought*. London: Croom Helm, 38–60.

Sand, G. 1861: *Le marquis de Villemer*. Paris: Lévy.

Sarre, P. 1987: Realism in practice, *Area*, 19, 3–10.

Saunders P. and Williams, P.R. 1986: The new conservatism: some thoughts on recent and future developments in urban studies, *Environment and Planning D: Society and Space*, 4, 393–9.

Sayer, R.A. 1982a: Explaining manufacturing shift: a reply to Keeble, *Environment and Planning A*, 14, 119–25.

— 1982b: Explanation in economic geography, *Progress in Human Geography*, 6, 68–88.

— 1984: *Method in social science: a realist approach*. London: Hutchinson.

— 1985a: Industry and space: a sympathetic critique of radical research, *Environment and Planning D: Society and Space*, 3, 3–29.

— 1985b: The difference that space makes. In D. Gregory and J. Urry (eds.), *Social relations and spatial structures*. London: Macmillan, 49–66.

— 1988: Some misconceptions about realism in geography. Paper delivered to the *Annual Conference, Institute of British Geographers*, Loughborough, 7 Jan.

— and Morgan, K. 1985: A modern industry in a declining region, links between method, theory and policy. In D. Massey and R. Meegan (eds.), *Politics and method: contrasting studies in industrial geography*. London: Methuen, 147–68.

Schaefer, F. K. 1953: Exceptionalism in geography: a methodological examination, *Annals of the Association of American Geographers*, 43, 226–49.

Scheffler, I. 1974: *Four pragmatists: a critical introduction to Peirce, James, Mead, and Dewey*. London: Routledge and Kegan Paul.

Schilpp, P. A. (ed.) 1974: *The philosophy of Karl Popper*, 2 vols. Library of Living Philosophers; La Salle, Ind.: Open Court. (Vol. i contains a miscellany of essays on the work of the philosopher; in vol. ii he replies, see Popper, 1974.)

Scholes, P. A. 1942: *The Oxford companion to music*. Oxford: Oxford University Press.

Schumm, S. A. 1979: Geomorphic thresholds, *Transactions of the Institute of British Geographers*, 4, 485–515.

— and Lichty, R. W. 1965: Time, space and causality in geomorphology, *American Journal of Science*, 263, 110–19.

Schutz, A. 1962: *Collected papers*, i and ii. The Hague: Martinus Nijhoff.

— 1967: *The phenomenology of the social world*. Evanston: Northwestern Univeristy Press.

— and Luckmann, T. 1974: *The structures of the life-world*. London: Heinemann.

Scott, A. J. and Roweis, S. T. 1977: Urban planning in theory and practice: a reappraisal, *Environment and Planning A*, 9, 1097–119.

Scott, P. 1959: The Australian CBD, *Economic Geography*, 35, 290–314.

Seamon, D. 1979: *A geography of the lifeworld*. New York: St Martin's.

— 1984: The question of reliable knowledge: the irony and tragedy of positivist research, *The Professional Geographer*, 36, 216–18.

Sellier, P. (ed.) 1976: *Pensées de Blaise Pascal*. Paris: Mercure de France.

Senior, M. L. 1979: From gravity modelling to entropy maximising: a pedagogic guide, *Progress in Human Geography*, 3, 175–210.

Shepard, R. N. 1978: The mental image, *American Psychologist*, 33, 125–37.

Shoard, M. 1982: The lure of the moors. In J. R. Gold and J. Burgess (eds.), *Valued environments*. London: Allen and Unwin, 55–73.

Short, J. R. 1982: *An introduction to political geography*. London: Routledge and Kegan Paul.

Simon, H. A. 1952: A behavioral model of rational choice, *Quarterly Journal of Economics*, 69, 99–118.

— 1957: *Models of man: social and rational*. New York: Wiley.

Simpson, G. G. 1963: Historical science. In C. C. Albritton (ed.), *The fabric of geology*. Reading, Mass.: Addison-Wesley, 24–48.

Slater, D. 1977: The poverty of modern geographical enquiry. In R. Peet (ed.), *Radical geography*. London: Methuen, 40–57.

Slaymaker, H. O. 1968: Scale problems in hydrology. In E. G. Bowen and J. A. Taylor (eds.), *Geography at Aberystwyth*. Cardiff: University of Wales Press, 68–86.

Smailes, A. E. 1971: Urban systems, *Transactions and Papers of the Institute of British Geographers*, 53, 1–14.

Small, R. J. 1980: The Tertiary evolution of south-east England: an alternative interpretation. In D.K.C. Jones (ed.), *The shaping of southern England*. London: Academic Press, 49–70.

Smalley, I.J. and Vita-Finzi, C. 1969: The concept of 'system' in the earth sciences, particularly geomorphology, *Bulletin of the Geological Society of America*, 80, 1591–4.

Smith, C. 1983: A case study of structuralism: the pure bred beef business, *Journal for Theory of Social Behaviour*, 13, 3–18.

Smith. D.M. 1972: *Industrial location*. New York: Wiley.

—— 1977: *Human geography: a welfare approach*. London: Arnold.

—— 1981a: Marxian economics. In R.J. Johnston, D. Gregory, P. Haggett, D. Smith, and D.R. Stoddart (eds.), *A dictionary of human geography*. Oxford: Blackwell, 203–9.

—— 1981b: Welfare geography. In R.J. Johnston, D. Gregory, P. Haggett, D. Smith, and D.R. Stoddart (eds.), *A dictionary of human geography*. Oxford: Blackwell, 370–2.

—— 1986:UGC research ratings: pass or fail? *Area*, 18, 247–50.

Smith, N. 1979: Geography, science and post-positivist modes of explanation, *Progress in Human Geography*, 3, 356–83.

—— 1984: *Uneven development: nature, capital and the production of space*. Oxford: Blackwell.

—— 1987: Dangers of the empirical turn: some comments on the CURS initiative, *Antipode*, 19, 59–68.

Smith, S.J. 1984: Practicing humanistic geography, *Annals of the Association of American Geographers*, 74, 353–74.

Smith, T.L. 1937: *The sociology of rural life*. New York: Harper.

Soja, E.W. 1980: The socio-spatial dialectic, *Annals of the Association of American Geographers*, 70, 207–25.

—— 1985: The spatiality of social life: towards a transformative retheorisation. In D. Gregory and J. Urry (eds.), *Social relations and spatial structure*. London: Macmillan, 90–127.

—— 1987: The postmodernization of geography: a review, *Annals of the Association of American Geographers*, 77, 289–94.

Sorell, T. 1990: The world from its own point of view. In A. Malachowski, (ed.), *Reading Rorty*, 11–25.

Spengler, O. 1926: *The decline of the west*. New York: Knopf.

Stafford, H. 1974: The anatomy of the location decision: content analysis of case studies. In F.E.I. Hamilton (ed.), *Spatial perspectives on industrial organization and decision-making*. London: Wiley, 169–87.

Steinbeck, J. 1939: *The grapes of wrath*. London: Heinemann.

Stewart, J.Q. 1945: *Coasts, waves and weather for navigators*. Boston: Ginn.

—— and Warntz W. 1958: Macrogeography and social science, *Geographical Review*, 48, 167–84.

Stoddart, D.R. 1966: Darwin's impact on geography, *Annals of the Association of American Geographers*, 56, 683–98.

—— 1967: Organism and ecosystem as geographical models. In R.J. Chorley and P. Haggett (eds.), *Models in geography*. London: Methuen, 511–48.

—— 1987: To claim the high ground: geography for the end of the century, *Transactions of the Institute of British Geographers*, 12, 327–37.

Storper, M. 1985: The spatial and temporal constitution of social action: a critical reading of Giddens, *Environment and Planning D: Society and Space*, 3, 407–24.

—— 1987: The post-Enlightenment challenge to Marxist urban studies, *Environment and Planning D: Society and Space*, 5, 418–26.

Stove, D. C. 1982: *Popper and after: four modern irrationalists*. Oxford: Pergamon.

Sullivan, H. S. 1953: *The interpersonal theory of psychiatry*. London: Tavistock.

Svart, L. 1974: On the priority of behaviour in behavioural research: a dissenting view, *Area*, 6, 301–5.

Tansley, A. G. 1935: The use and abuse of vegetational concepts and terms, *Ecology*, 16, 284–307.

Taylor, C. 1987: Overcoming epistemology. In K. Baynes, J. Bohman, and T. McCarthy (eds.), *After philosophy: end or transformation?* Cambridge, Mass.: MIT Press, 464–88.

Taylor, L. 1990: Rorty in the epistemological tradition. In Malachowski, A. (ed.), *Reading Rorty*, 257–75.

Taylor, P. J. 1975: *Distance decay models in spatial interactions*. Concepts and Techniques in Modern Geography (Catmog), No. 2. Norwich: Geo Abstracts.

—— 1976: An interpretation of the quantification debate in British geography, *Transactions of the Institute of British Geographers*, NS1, 129–42.

—— 1982: A materialist framework for political geography, *Transactions of the Institute of British Geographers*, 7, 14–34.

—— 1985: *Political geography: world-economy, nation state and community*. London: Longman.

—— 1987: The paradox of geographical scale in Marx's politics, *Antipode*, 19, 287–306.

Teilhard de Chardin, P. 1959: *The phenomenon of man*. London: Collins.

Thayer, H. S. 1967: Pragmatism, *The encyclopaedia of philosophy*, New York: Collier Macmillan, vi. 430–6.

Theocharis, T. and Psimopoulos, M. 1987: Where science has gone wrong, *Nature*, 329, 595–8.

Thompson, M. 1978: Review of N. Rescher, *Methodological pragmatism*, *Philosophy of Science*, 45, 493–5.

Thornes, J. B. 1987: Environmental systems: patterns, processes and evolution. In M. J. Clark, K. J. Gregory, and A. M. Gurnell (eds.), *Horizons in physical geography*. London: Macmillan, 27–46.

—— and Ferguson, R. I. 1981: Geomorphology. In N. Wrigley and R. J. Bennett (eds.), *Quantitative geography: a British view*. London: Routledge and Kegan Paul, 284–93.

Thrift, N. 1981: Behavioural geography. In N. Wrigley and R. J. Bennett (eds.), *Quantitative geography: a British view*. London: Routledge and Kegan Paul, 352–65.

—— 1983: On the determination of social action in space and time. *Environment and Planning D: Society and Space*, 1, 23–57.

—— 1987: No perfect symmetry, *Environment and Planning D: Society and Space*, 5, 400–7.

284   References

—— and Pred, A. 1981: Time geography: a new beginning, *Progress in Human Geography*, 5, 277–86.

Tiebout, C.M. 1957: Location theory, empirical evidence, and economic evolution, *Papers and Proceedings of the Regional Science Association*, 3, 74–86.

Tinkler, K.J. 1977: *An introduction to graph theoretical methods in geography*. Concepts and techniques in modern geography (Catmog), No. 14. Norwich: Geo Abstracts.

—— 1985: *A short history of geomorphology*. London: Croom Helm.

Tolstoy, L.N. 1957: Epilogue, part II. *War and Peace*, Trans. R. Edmonds; Harmondsworth: Penguin, ii. 1400–44.

Trusov, Y. 1969: The concept of the noosphere, *Soviet Geography*, 10, 220–36.

Tuan, Yi-fu, 1971: Geography, phenomenology and the study of human nature, *The Canadian Geographer*, 15, 181–92.

—— 1973: Ambiguity in attitudes towards environment, *Annals of the Association of American Geographers*, 63, 411–23.

—— 1974*a*: *Topophilia: a study of environmental perception, attitudes and values*. Englewood Cliffs, NJ: Prentice-Hall.

—— 1974*b*: Space and place: humanistic perspective, *Progress in Human Geography*, 6, 211–52.

—— 1976: Humanistic geography, *Annals of the Association of American Geographers*, 66, 266–76.

—— 1978*a*: Literature and geography: implications for geographical research. In D. Ley and M. Samuels (eds.), *Humanistic geography: prospects and problems*. London: Croom Helm, 194–206.

—— 1978*b*: Sign and metaphor, *Annals of the Association of American Geographers*, 68, 363–72.

Turner, F.J. 1894: The significance of the frontier in American history, *Annual Report of the American Historical Association for 1893*. Washington: US Government Printing Office.

Tyler, S.A. 1978: *The said and the unsaid*. London: Academic Press.

Ullman, E.L. 1967: Geographic prediction and theory: the measure of recreation benefits in the Meramec basin. In S.B. Cohen (ed.), *Problems and Trends in American Geography*. New York: Basic Books, 124–45.

Urry, J. 1985: Social relations, space and time. In D. Gregory and J. Urry (eds.), *Social relations and spatial structures*. London: Macmillan, 20–48.

Van der Laan, L. and Piersma, A. 1982: The image of man: paradigmatic cornerstone in human geography, *Annals of the Association of American Geographers*, 72, 411–26.

Vernon, P.E. (ed.) 1970: *Creativity*. Harmondsworth: Penguin.

Vickers, G. 1970: *Freedom in a rocking boat*. London: Allen Lane.

Von Bertalanffy, L. 1968: *General systems theory: foundation, development, applications*. New York: Braziller.

Waddell, E. 1977: The hazards of scientism: a review article, *Human Ecology*, 5, 69–76.

Wagstaff, J.M. 1978: A possible interpretation of settlement pattern evolution in terms of 'Catastrophe Theory', *Transactions of the Institute of British Geographers*, 4, 438–44.

—— 1979: Dialectical materialism, geography and catastrophe theory, *Area*, 11, 326–32.

—— 1980: Dialectical materialism and geography. (Reply to M.D. Day), *Area*, 12, 146–9.

Wallas, G. 1926: *The art of thought*. New York: Harcourt Brace.

Wallerstein, I. 1975: Class formation in a capitalist world-economy, *Politics and Society*, 5, 367–75.

—— 1984: *Politics of world economy*. Cambridge: Cambridge University Press.

Walmsley, D.J. 1974: Positivism and phenomenology in human geography, *The Canadian Geographer*, 18, 95–107.

—— and Lewis, G.J. 1984: *Human geography: behavioural approaches*. London: Longman.

—— and Sorenson, A.D. 1980: What Marx for the radicals? An antipodean viewpoint, *Area*, 137–41.

Warntz, W. 1973: New geography as general spatial systems theory: old social physics writ large. In R.J. Chorley (ed.), *Directions in geography*, London: Methuen, 89–126.

—— 1984: Trajectories and co-ordinates. In M. Billinge, D. Gregory, and R. Martin (eds.), *Recollections of a revolution: geography as spatial science*. London: Macmillan, 134–50.

Watkins, J.W.N. 1970: Against 'normal science'. In I. Lakatos and A. Musgrave (eds.), *Criticism and the growth of knowledge*. Cambridge: Cambridge University Press, 25–38.

Watson, M.K. 1978: The scale problem in human geography, *Geografiska Annaler*, 60B, 36–47.

Watts, S.J. and Watts, S.J. 1978: On the idealist alternative in geography and history, *The Professional Geographer*, 30, 123–7.

Weaver, W. 1967: *A quarter century in the natural sciences: Annual Report of the Rockefeller Foundation*. First pub. 1958; New York, 7–122.

—— 1967: Science and complexity, *Science and imagination*. New York: Basic Books, 25–33.

Webb, J.W. 1976: Geographers and scales. In L.A. Kosinski and J.W. Webb (eds.), *Population at microscale*. Hamilton, NZ: International Geographical Union Commission on Population Geography and New Zealand Geographical Society, 13–19.

Webber, M.J. 1972: *Impact of uncertainty on location*. Canberra: Australian National University Press.

—— 1977: Pedagogy again: what is entropy? *Annals of the Association of American Geographers*, 67, 254–66.

Weber, A. 1909: Über den Standorf der Industrien. In *Alfred Weber's theory of location*. Trans. C.J. Friedrich; Chicago: University of Chicago Press.

Welch, R.V. 1978: Conflict, apparent and real: a review of concepts of space in geography, *New Zealand Geographer*, 34, 24–30.

Wheatley, P. 1971: *The pivot of the four quarters: a preliminary enquiry into the origins and character of the ancient Chinese city*. Chicago: Aldine.

Wheelwright, P. 1954: *The burning fountain: a study in the language of symbolism*. Bloomington, Ind.: Indiana University.

White, G. F. 1961: The choice of use in reserve management, *Natural Resources Journal*, 1, 23–40.

Whittaker, R. H. 1953: A consideration of climax theory: the climax as a population and pattern, *Ecological Monographs*, 23, 41–78.

Williams, B. 1985: *Ethics and the limits of philosophy*. London: Fontana.

— 1990: Auto-da-fe: consequences of pragmatism. In Malachowski, A. (ed.), *Reading Rorty*, 26–37.

Williams, M. 1987: Editorial: can the centre hold? *Transactions of the Institute of British Geographers*, 12, 387–90.

Wilson, A. G. 1972: Theoretical geography, *Transactions of the Institute of British Geographers*, 57, 31–44.

— 1974: *Urban and regional models in geography and planning*. Chichester: Wiley.

— 1981: *Geography and the environment: systems analytical methods*. Chichester: Wiley.

Wirth, L. 1938: Urbanism as a way of life, *American Journal of Sociology*, 44, 1–24.

Wisdom, J. O. 1966: The need for corroboration, *Technology and Culture*, 7, 367–70.

— 1987: *Philosophy of the social sciences*, i: *A metascientific introduction*; ii: *Schemata*. Aldershot: Gower.

Wolpert, J. 1964: The decision process in spatial context, *Annals of the Association of American Geographers*, 54, 337–58.

— 1965: Behavioral aspects of the decision to migrate, *Papers and Proceedings, Regional Science Association*, 15, 159–72.

— 1970: Departures from the usual environment in locational analysis, *Annals of the Association of American Geographers*, 60, 220–9.

Women on the move 1985–7: London: London Strategic Policy Unit.

Wood, J. D. 1982: Rethinking geographical inquiry: prologue, chorus, epilogue. In J. D. Wood (ed.), *Rethinking geographical inquiry*. Geographical Monographs, No. 11. York: York University, 1–27.

Wooldridge, S. W. and East, W. G. 1958: *The spirit and purpose of geography*. London: Methuen.

— and Linton, D. L. 1955: *Structure, surface and drainage of south-east England*. London: Philip.

— and Morgan, R. S. 1937: *The physical basis of geography*. London: Longman.

Young, O. R. 1964: A survey of general systems theory, *General Systems*, 9, 61–80.

Zeigler, B. P. 1976: *Theory of modelling and simulation*. New York: Wiley.

Zelinsky, W. 1975: The demigod's dilemma, *Annals of the Association of American Geographers*, 65, 123–43.

Zukav, G. 1980: *The dancing Wu Li masters*. London: Fontana.

AUTHOR INDEX

(Compiled by O. J. Bird)

A page number followed by n. refers to a source or note to either a figure or a table.

SUBJECT INDEX

Preliminary alphabetical list. It may be useful to provide here a list of all those entries in the Subject Index which end in -ism.

Bold-face entries indicate major references.
Italicized page numbers refer to figures.
A page number followed by n. refers to a source or note to either a figure or a table.